W0227659

CAMBRIDGE MONOGRAPHS
ON MECHANICS AND APPLIED MATHEMATICS

GENERAL EDITORS

G. K. BATCHELOR, PH.D., F.R.S.
Lecturer in Mathematics in the University of Cambridge

H. BONDI, M.A.
Professor of Applied Mathematics at King's College, University of London

SOUND PULSES

SOUND PULSES

BY

F. G. FRIEDLANDER, PH.D.
Lecturer in Mathematics in the University of Cambridge

CAMBRIDGE
AT THE UNIVERSITY PRESS
1958

CAMBRIDGE UNIVERSITY PRESS
Cambridge, New York, Melbourne, Madrid, Cape Town, Singapore, São Paulo, Delhi

Cambridge University Press
The Edinburgh Building, Cambridge CB2 8RU, UK

Published in the United States of America by Cambridge University Press, New York

www.cambridge.org
Information on this title: www.cambridge.org/9780521117500

First published 1958
This digitally printed version 2009

A catalogue record for this publication is available from the British Library

ISBN 978-0-521-05039-5 hardback
ISBN 978-0-521-11750-0 paperback

CONTENTS

4 The Application of Geometrical Acoustics to Reflexion Problems

5 The Diffraction of a Pulse by a Wedge

6 Some other Diffraction Problems

PREFACE

The theory of sound is a long-established branch of mathematical physics. Its principal results are well known, and there are many text-books devoted to it. These deal in the main with periodic disturbances, such as harmonic wave trains and standing waves. Aperiodic disturbances receive much less attention; it is usually pointed out that they can be resolved into harmonic components by Fourier's integral theorem, and only the elementary examples of plane and spherical waves are treated directly. There is, however, a theory of aperiodic disturbances which is based on the theory of linear partial differential equations of hyperbolic type. It is particularly effective in the case of a disturbance with a clearly defined front; such a disturbance may be called a sound pulse.† There does not seem to be an adequate description of the theory of sound pulses and of its recent developments. The present monograph is an attempt to fill this gap; but it is not a complete survey of the field, since the selection of the material was governed to some extent by personal taste.

The need for an independent treatment of pulses may be illustrated by the following example. It is well known that pulse fronts are wave fronts in the sense of geometrical optics; they are propagated, reflected, refracted and diffracted in accordance with Fermat's principle. To establish this by means of Fourier analysis would require a profound study of the behaviour of harmonic waves as the frequency tends to infinity. But from the point of view of the theory of hyperbolic equations, it is merely another way of stating that pulse fronts, considered as hypersurfaces in space-time, are characteristics of the wave equation. This was already emphasized by Hadamard in his *Leçons sur la propagation des ondes* in 1903. Hadamard based his argument on the Cauchy-Kowalewsky existence theorem which is valid for analytic solutions only, and of limited applicability. But since then, a simple

† The term pulse also suggests that the disturbance should end abruptly. This is in general not possible, except for plane and spherical pulses in a homogeneous medium. A plane pulse of finite duration reflected at an obstacle gives rise (in general) to a secondary pulse with a definite beginning, but with a 'tail' of infinite duration.

and elementary proof has been discovered independently no less than three times, by Zaremba in 1915, by Rubinowicz in 1920, and by Friedrichs and Lewy in 1928. It is nevertheless still not as widely known among applied mathematicians as it deserves to be.

This monograph is based throughout on the scalar wave equation of acoustics, which is a linear approximation that can only be applied to small disturbances. The derivation of higher approximations, that take some of the additional features due to the non-linearity of the exact equations of motion of a fluid into account, is an important and difficult subject. I have not attempted to deal with it here because I do not feel competent to do so; the reader who wants to pursue this matter further will find an excellent account by Professor M. J. Lighthill in the recently published *General theory of high speed aerodynamics*, edited by W. Sears (Princeton 1954).

Once the acoustic approximation is adopted, the development of the subject becomes a mathematical rather than a physical investigation. This book is therefore essentially an essay on the pulse solutions of the wave equation. The mathematical problems are in many respects similar to those encountered in optics and the theory of electromagnetic waves, and I hope that workers in these fields may also find the book of some interest, although it treats the acoustic case.

Both in developing the general theory of 'geometrical acoustics' in chapter 3, and in the solution of particular problems, I have used the *theory of distributions* of Laurent Schwartz. I feel sure that this theory, or one of its variants, such as the 'generalized functions' of Mikusinsky whose use has been advocated by Temple, will in the immediate future become an accepted and useful part of the technical equipment of mathematical physicists. As it happens, only the elements of the theory are, for the most part, required here, and I hope that no one will be deterred by its comparative unfamiliarity. Apart from technical advantages, a case can be made for the adoption of the theory of distributions in any classical physical theory based on linear equations, as I have attempted to show at the end of the introduction to chapter 3.

Some of the work has not been published before; this includes the appendix to chapter 3 on the focusing of acoustic shocks, the method by which the Green's function of the wedge is derived in chapter 5, and the discussion of pulse diffraction by a sphere in chapter 6.

I should like to thank Professor H. G. Garnir, who has read the manuscript. I also want to express my appreciation of the many interesting discussions with my colleagues at the Institute of Mathematical Sciences, New York University, and particularly the members of the Electromagnetics Group, during a six months' visit in 1954.† These have proved most helpful during the subsequent writing of this book.

F. G. F.

CAMBRIDGE
1957

† This visit was made possible by the sponsorship of the U.S. Air Force Cambridge Research Center.

CHAPTER I

INTRODUCTION

1. Sound pulses

The theory of sound deals with the propagation of small disturbances in a fluid. The classical theory, developed in the nineteenth century, is chiefly concerned with disturbances that vary periodically and can be analysed into wave trains or standing waves for which the pressure, density and particle velocity vary sinusoidally with time. Any aperiodic disturbance, which for shortness we may call a sound pulse, can of course be represented as the result of the superposition of harmonic wave trains; but there are many aspects of the theory of sound pulses which can more conveniently or more concisely be considered directly without having resort to Fourier analysis. Certain quite basic properties of sound pulses, such as the fact that disturbance fronts propagate according to the laws of geometrical optics and that the pressure rise at an (acoustic) shock wave varies along a ray according to the intensity law of geometrical optics, are examples of results which can be obtained in this way. The mathematical background of this approach is the theory of linear hyperbolic partial differential equations of the second order; this theory has largely been developed since the turn of the century, and has still not reached a final form. In addition to this, a great deal can be achieved by Laplace transform methods, which again post-date the development of classical acoustics; it is therefore not surprising that the treatment of sound pulses, as distinct from harmonic sound waves, is a more recent development.

The theory of sound pulses is in the same relation to that part of gas dynamics which deals with unsteady motions, and particularly waves of finite amplitude, as the 'linearized theory of supersonic flow' is to the non-linear theory of the steady flow of a compressible fluid. In fact, these two subjects are closely related to each other, and because of the technological importance of steady flow the linearized theory of supersonic flow has received much more attention and has a far more extensive literature than that of sound pulses. There seems to be, however, a good case for the theory of sound pulses to be treated as a separate subject; on the one hand it completes the classical theory of acoustics, on the other hand it can serve as an approximate treatment of the much more complex subject of unsteady gas dynamics, whose problems are practically intractable, unless symmetry reduces the number of spatial co-ordinates to one. Also, it is to be hoped that the efforts which have already been made to bridge the gap between the linearized and the full theory, by developing methods which use the results of linearized theory as the starting point for the calculation of better approximations, will continue and in time become a recognized and useful part of gas dynamics; the theory of sound pulses would then be an essential prolegomenon to such a theory.

The approximations which are made to obtain the equations of motion of the theory of sound are of two kinds. The first, which are also often made in gas dynamics, consist in neglecting the two transport phenomena of viscosity and heat conduction. The nature of this approximation is discussed in the general theory of gas dynamics and there is no need to go into it here. The only point to which we must draw attention is the well-known one that, when transport phenomena are ignored, certain discontinuous transitions must be admitted. These are of two kinds: shock waves and contact discontinuities. Shock waves are instantaneous compression fronts propagated with a speed exceeding that of sound in the medium ahead of the wave. Contact discontinuities (or vortex sheets) are 'material surfaces' swept along with the flow across which the tangential particle velocity, the density and the entropy are discontinuous. Analogous discontinuities must be taken into account in the theory of sound pulses; they will be called acoustic shocks, and acoustic contact discontinuities.

The approximation which characterizes the theory of sound is based on the idea that if the disturbances to be considered are small, then the non-linear terms in the equation of motions of an inviscid, non-heat-conducting fluid can be neglected. What one is aiming at, of course, are approximate solutions, and it is obviously not a foregone conclusion that these can be obtained by approximating the differential equations of motion. In fact, the (absolute) error in the solution is of the order of the product of the distance through which a pulse is propagated and of the ratio of the terms neglected in the equations of motion to those retained. This is the most serious drawback in the application of the acoustic approximation; several methods for the improvement of the theory in this respect have been proposed, both with regard to problems of unsteady and of steady flow. An account of these has been recently given by M. J. Lighthill.†

2. The equations of motion

This book is concerned exclusively with the approximate acoustic equations of motion and the wave equation which is derived from them. Consider the propagation of small disturbances in a medium on which no body force acts but which may be inhomogeneous when it is at rest. The conditions of hydrostatic equilibrium then imply that the pressure p_0 of the medium at rest is constant. The density ρ_0 and the specific entropy S_0 may be functions of position. They are not independent of each other since p_0, ρ_0 and S_0 are connected by the equation of state.

Now let p_0+p and $\rho_0+\rho$ be the pressure and density respectively at the point $\mathbf{x}$ at time t when the medium is disturbed, and let $\mathbf{u}$ be the vector particle velocity.‡ Then the exact equation of conservation of mass is

$$\frac{\partial}{\partial t}(\rho_0+\rho)+\operatorname{div}[(\rho_0+\rho)\,\mathbf{u}]=0, \qquad (1.2.1)$$

and the exact momentum equations for an inviscid medium are

$$(\rho_0+\rho)\left\{\frac{\partial \mathbf{u}}{\partial t}+(\mathbf{u}\,.\operatorname{grad})\,\mathbf{u}\right\}+\operatorname{grad}(p_0+p)=0. \qquad (1.2.2)$$

If we note that p_0 is constant and that ρ_0 is independent of t, and retain only the linear terms in these equations, then we obtain the corresponding acoustic equations

$$\frac{\partial \rho}{\partial t}+\operatorname{div}(\rho_0\mathbf{u})=0, \qquad (1.2.3)$$

$$\rho_0\frac{\partial \mathbf{u}}{\partial t}+\operatorname{grad} p=0. \qquad (1.2.4)$$

† Lighthill (1954).

‡ The components of the position vector $\mathbf{x}$ are the coordinates x_1, x_2, x_3 and those of $\mathbf{u}$ are u_1, u_2, u_3. We shall use both vector and scalar notation; the summation convention will be employed in the form that a repeated Greek suffix is to be summed from 1 to 3.

In addition to (1.2.1) and (1.2.2) the energy equation, that is to say, the first law of thermodynamics, must hold. It is well known that except at a discontinuity front this implies that the specific entropy S_0+S is carried with the fluid, that is to say,

$$\frac{\partial}{\partial t}(S_0+S)+\mathbf{u}\,.\,\mathrm{grad}\,(S_0+S)=0. \qquad (1.2.5)$$

Since S_0 is independent of t the equivalent acoustic equation is

$$\frac{\partial S}{\partial t}+\mathbf{u}\,.\,\mathrm{grad}\,S_0=0. \qquad (1.2.6)$$

When a particle passes through a shock front its entropy changes abruptly. But the entropy change is of the third order in the shock strength,† so that the law of entropy conservation applies in the acoustic approximation even when shocks are present. We shall make use of this fact in chapter 3 where the equations of motion will be reinterpreted in such a manner that they apply to discontinuous as well as to continuous disturbances. It will then be legitimate to work with (1.2.6). The five differential equations of motion, (1.2.3), (1.2.4) and (1.2.6), contain the six dependent variables p, ρ, S, $\mathbf{u}$, and so a further relation is required to obtain a determinate system. This is the linearized equation of state,

$$p=\frac{\partial p_0}{\partial \rho_0}\rho+\frac{\partial p_0}{\partial S_0}S=c_0^2\rho+A_0S, \qquad (1.2.7)$$

where $c_0=(\partial p_0/\partial \rho_0)^{\frac{1}{2}}$ is the velocity of sound in the medium at rest and $A_0=\partial p_0/\partial S_0$.

We can now eliminate ρ and S as follows. By (1.2.6) and (1.2.7)

$$\frac{\partial p}{\partial t}=c_0^2\frac{\partial \rho}{\partial t}+A_0\frac{\partial S}{\partial t}=c_0^2\frac{\partial \rho}{\partial t}-A_0\mathbf{u}\,.\,\mathrm{grad}\,S_0. \qquad (1.2.8)$$

Since p_0 is constant,

$$0=\mathrm{grad}\,p_0=c_0^2\,\mathrm{grad}\,\rho_0+A_0\,\mathrm{grad}\,S_0.$$

Hence $-A_0\mathbf{u}\,.\,\mathrm{grad}\,S_0=c_0^2\mathbf{u}\,.\,\mathrm{grad}\,\rho_0$, so that (1.2.8) becomes

$$\frac{\partial p}{\partial t}=c_0^2\left(\frac{\partial \rho}{\partial t}+\mathbf{u}\,.\,\mathrm{grad}\,\rho_0\right),$$

† Cf. Courant and Friedrichs (1948), p. 156.

which by (1.2.3) is
$$\frac{1}{\rho_0 c_0^2}\frac{\partial p}{\partial t}+\operatorname{div}\mathbf{u}=0. \tag{1.2.9}$$

The equations (1.2.4) and (1.2.9) which contain only p and $\mathbf{u}$ are a convenient form of the acoustic equations of motion. If ρ and S are also required they can be determined from (1.2.3) and (1.2.7) once p and $\mathbf{u}$ are known.†

3. The wave equation

The most important dependent variable from a physical point of view is usually the excess pressure p. The equation satisfied by p can be obtained at once by eliminating $\mathbf{u}$ between (1.2.4) and (1.2.9):
$$\frac{1}{\rho_0 c_0^2}\frac{\partial^2 p}{\partial t^2}-\operatorname{div}\left(\frac{1}{\rho_0}\operatorname{grad} p\right)=0. \tag{1.3.1}$$

This is a generalized wave equation; we shall refer to it simply as the wave equation from now on. If the medium is homogeneous when at rest so that ρ_0 and c_0 are constant then (1.3.1) reduces to the ordinary wave equation
$$\frac{1}{c_0^2}\frac{\partial^2 p}{\partial t^2}-\nabla^2 p=0. \tag{1.3.2}$$

The equations of motion (1.2.4) and (1.2.9) imply the wave equation. It is natural to ask what velocity fields $\mathbf{u}$ can be associated with a given solution p of the wave equation. This question can be answered as follows. Let us specify in addition to p an initial velocity field $\mathbf{u}_0=\mathbf{u}_0(\mathbf{x})$. Then we must define $\mathbf{u}$ in accordance with (1.2.4) by
$$\rho_0\mathbf{u}=\rho_0\mathbf{u}_0-\int_0^t \operatorname{grad} p\,\mathrm{d}t. \tag{1.3.3}$$

It remains to satisfy (1.2.9). Now
$$\begin{aligned}\operatorname{div}\mathbf{u}&=\operatorname{div}\mathbf{u}_0-\int_0^t \operatorname{div}\left(\frac{1}{\rho_0}\operatorname{grad} p\right)\mathrm{d}t\\ &=\operatorname{div}\mathbf{u}_0-\int_0^t \frac{1}{\rho_0 c_0^2}\frac{\partial^2 p}{\partial t^2}\,\mathrm{d}t,\end{aligned}$$

whence
$$\operatorname{div}\mathbf{u}+\frac{1}{\rho_0 c_0^2}\frac{\partial p}{\partial t}=\operatorname{div}\mathbf{u}_0+\frac{1}{\rho_0 c_0^2}\left[\frac{\partial p}{\partial t}\right]_{t=0}.$$

† For the more general case of the propagation of sound in a moving medium, see Blokhintzev (1946).

Hence the equation (1.2.9) is satisfied provided that the initial velocity field is such that

$$\operatorname{div} \mathbf{u}_0 = -\frac{1}{\rho_0 c_0^2}\left[\frac{\partial p}{\partial t}\right]_{t=0}. \tag{1.3.4}$$

Apart from this condition, $\mathbf{u}_0$ is arbitrary. Hence two velocity fields which can be associated with the same solution p of the wave equation differ by a steady divergence-free field; such a field, together with $p=0$, satisfies the equations of motion. This indeterminacy of the velocity field is, in practice, of little importance.

It is sometimes more convenient to work with a velocity potential Φ rather than with p. It follows from (1.2.4) that the curl of the mass-flow vector $\rho_0 \mathbf{u}$ is independent of time. In many pulse problems the field vanishes for sufficiently large negative t, or is the sum of such a field and of a given field. In that case $\rho_0 \mathbf{u}$ is irrotational and one can put

$$\rho_0 \mathbf{u} = -\operatorname{grad}\Phi, \quad p = \frac{\partial \Phi}{\partial t}. \tag{1.3.5}$$

Then (1.2.4) holds automatically and (1.2.9) shows that ϕ also satisfies the wave equation (1.3.1),

$$\frac{1}{\rho_0 c_0^2}\frac{\partial^2 \Phi}{\partial t^2} - \operatorname{div}\left(\frac{1}{\rho_0}\operatorname{grad}\Phi\right) = 0. \tag{1.3.6}$$

4. The effect of body forces

When the medium is subject to body force then the acoustic equations of motion cannot be reduced to a single second-order equation like (1.3.1). But for short pulses an equation of this form still holds approximately, with the rest density ρ_0 replaced by a certain fictitious density, provided that the body force is derived from a potential.

Consider first the equilibrium state of the medium. If U is the body-force potential per unit mass then the pressure p_0 must satisfy the equilibrium conditions

$$\operatorname{grad} p_0 = -\rho_0 \operatorname{grad} U. \tag{1.4.1}$$

This equation implies that both p_0 and ρ_0 are constant on the equipotentials, so that $p_0 = p_0(U)$ and $\rho_0 = \rho_0(U)$. Hence also $S_0 = S_0(U)$. One therefore has two relations between these three variables: the equilibrium condition $dp_0/dU = -\rho_0$ and the equation of state. The values of a suitable function of ρ_0 and S_0, for instance the temperature, can be prescribed arbitrarily for each equipotential.

The momentum equations (1.2.4) must now be replaced by

$$\rho_0 \frac{\partial \mathbf{u}}{\partial t} + \operatorname{grad} p = -\rho \operatorname{grad} U. \qquad (1.4.2)$$

But for short pulses the term $-\rho \operatorname{grad} U$ can be neglected. In fact, ρ is of the order of p/c_0^2 so that the magnitude of this term is of the order of pg/c_0^2, where p is a typical value of the excess pressure and g a typical value of $|\operatorname{grad} U|$. On the other hand, $\operatorname{grad} p$ is of the order of p/L, where L is a length characteristic of the disturbance propagated. If gL/c_0^2 is small, then the term $\rho \operatorname{grad} U$ can be neglected. For air, with gravity as the body force, c_0^2/g is of the order of 11 km., and c_0/g of the order of 34 sec.

We can therefore retain (1.2.3) and (1.2.4). Instead of (1.2.5) we can use the equivalent equation

$$\frac{\partial}{\partial t}(p_0+p) + \mathbf{u} . \operatorname{grad}(p_0+p) = c^2\left\{\frac{\partial}{\partial t}(\rho_0+\rho) + \mathbf{u} . \operatorname{grad}(\rho_0+\rho)\right\},$$

where c is the local velocity of sound; its acoustic approximation is

$$\frac{\partial p}{\partial t} + \mathbf{u} . \operatorname{grad} p_0 = c_0^2\left(\frac{\partial \rho}{\partial t} + \mathbf{u} . \operatorname{grad} \rho_0\right) = -\rho_0 c_0^2 \operatorname{div} \mathbf{u}. \qquad (1.4.3)$$

Elimination of $\mathbf{u}$ between this and (1.2.4) leads to the equation satisfied by the excess pressure,

$$\frac{\partial^2 p}{\partial t^2} = c_0^2\left\{\nabla^2 p + \frac{\operatorname{grad} p_0 - c_0^2 \operatorname{grad} \rho_0}{\rho_0 c_0^2} \operatorname{grad} p\right\}. \qquad (1.4.4)$$

This equation is not of the form (1.3.1) as it stands. But as ρ_0 and S_0, and hence also c_0, are functions of U only, we can define a 'fictitious density' ρ^* which is also a function of U by the equation

$$\frac{d}{dU} \log \frac{\rho^*}{\rho_0} = -\frac{1}{\rho_0 c_0^2}\frac{dp_0}{dU} = \frac{1}{c_0^2}, \qquad (1.4.5)$$

which implies that

$$\operatorname{grad}\left(\log \frac{\rho^*}{\rho_0}\right) = -\frac{1}{\rho_0 c_0^2} \operatorname{grad} p_0.$$

Now (1.4.4) can be written as

$$\frac{1}{\rho^* c_0^2}\frac{\partial^2 p}{\partial t^2} - \operatorname{div}\left(\frac{1}{\rho^*} \operatorname{grad} p\right) = 0, \qquad (1.4.6)$$

and this is again a wave equation of the form (1.3.1). Note that (1.4.5) is equivalent to

$$\frac{1}{\rho^*}\frac{d\rho^*}{dU} = -\frac{A_0}{\rho_0 c_0^2}\frac{dS_0}{dU}. \tag{1.4.7}$$

Hence if S_0 is constant—which in the case of the atmosphere means that the medium is in convective equilibrium—then ρ^* is constant and (1.4.6) is formally identical with the ordinary wave equation (1.3.2) except that c_0 is not constant.†

5. Boundary conditions

The wave equation is a linear equation of hyperbolic type. For such an equation, the initial-value problem is the basic boundary-value problem. Suppose first that the medium is unlimited and that there are no internal boundaries. Then we can specify its state at one instant, say when $t=0$, by giving p and $\mathbf{u}$ as functions of $\mathbf{x}$. and ask for the values of these variables at time t. Mathematically it is of no importance whether t is positive or negative, but in physical problems one is usually concerned with the state of the medium subsequent to the initial instant, that is to say, for positive t. Since the initial values of $\mathbf{u}$ determine the initial value of $\partial p/\partial t$ by (1.2.9) the data for the initial-value problem of the wave equation are the initial values of p and of $\partial p/\partial t$.

The initial-value problem, or rather the more general Cauchy problem in which the data are specified for $t=t(\mathbf{x})$, has been investigated extensively from the analytical point of view. In physical applications, the more difficult 'mixed' problem which arises when reflectors are present is more important. Unfortunately, much less work seems to have been done on this. Suppose that the medium fills a domain S of space whose boundary $\mathscr{B}$ is the surface of one or of several reflectors. Then in addition to the initial values a boundary condition on $\mathscr{B}$ must be specified. If the reflectors are rigid and fixed, then this is simply that the particle velocity component normal to the reflector must vanish. Hence both the normal derivative of the velocity potential and of the pressure (which is the

† When the medium is an ideal gas, ρ^* may be called the 'potential density', by analogy with the concept of 'potential temperature' which is used in theoretical meteorology. It is the density which a small volume of fluid would acquire if it were compressed adiabatically to a standard pressure.

normal acceleration) must vanish on $\mathscr{B}$. This is the boundary condition usually employed in work on pulses; if a pulse is short enough most reflectors may be expected to behave as rigid and fixed bodies, at least during a time interval of the order of the pulse length. A more general boundary condition which allows to some extent for the inevitable mobility of the boundary is obtained by assuming that each element of the reflector acquires a velocity normal to the surface which is proportional to the pressure,

$$u_n = \lambda p. \qquad (1.5.1)$$

In terms of the velocity potential or the pressure this condition becomes

$$\frac{\partial \phi}{\partial n} + \rho_0 \lambda \frac{\partial \phi}{\partial t} = 0, \quad \frac{\partial p}{\partial n} + \rho_0 \lambda \frac{\partial p}{\partial t} = 0, \qquad (1.5.2)$$

where $\partial/\partial n$ denotes the derivative along the normal to $\mathscr{B}$ drawn away from S. This condition includes the fixed and rigid boundary ($\lambda = 0$) and the case of a free surface ($\lambda = \infty$).

Although any reflexion problem can be formulated in this way, it is more usual when boundaries are present to specify an incident pulse. This can in principle of course only be done in the 'physical space' S, but, for instance, when the medium is homogeneous one may speak of the incident pulse as the disturbance which obtains when the reflector is absent. The total disturbance is then the sum of the incident field and a secondary field. If p_i denotes the incident pressure pulse and p_s the additional pressure pulse due to reflexion at the obstacles, then, for instance, in the case of fixed rigid boundaries $\partial p_s/\partial n = -\partial p_i/\partial n$ on $\mathscr{B}$. The secondary field must also satisfy a further condition, whose exact formulation will be given in § 6 of chapter 2; its purpose is to ensure that it is a pulse 'coming from' the reflector. It is the pulse analogue of the radiation condition which is imposed in the case of harmonic wave trains.

6. Poisson's solution of the initial-value problem

From now on, we shall drop the suffix zero and denote the density and velocity of sound in the undisturbed fluid by ρ and c respectively. The wave equation in a homogeneous medium therefore assumes the usual form

$$\frac{1}{c^2}\frac{\partial^2 p}{\partial t^2} - \nabla^2 p = 0. \qquad (1.6.1)$$

The solution of the initial-value problem of this equation is classical; it is due to Poisson. Let

$$p = p_0(\mathbf{x}), \quad \frac{\partial p}{\partial t} = p_1(\mathbf{x}) \quad (t=0). \tag{1.6.2}$$

Then

$$p = tM_{ct}(p_1) + \frac{\partial}{\partial t}\{tM_{ct}(p_0)\}, \tag{1.6.3}$$

where, for instance, $M_{ct}(p_0)$ denotes the mean value of p_0 over the sphere of radius ct with centre $\mathbf{x}$,

$$M_{ct}(p_1) = \frac{1}{4\pi}\int_0^{\pi}\int_0^{2\pi} p_1(x_1 + ct\sin\alpha\cos\beta,$$
$$x_2 + ct\sin\alpha\sin\beta,\ x_3 + ct\cos\alpha)\sin\alpha\,\mathrm{d}\alpha\,\mathrm{d}\beta. \tag{1.6.4}$$

The proof of this result may be found in the literature,† but will be given here for the reader's convenience.

Let $\mathbf{y}$ be a fixed point and

$$\mathbf{R} = \mathbf{x} - \mathbf{y}. \tag{1.6.5}$$

Given any function $F(\mathbf{x}, t)$ one can define a function $[F]$ by

$$[F] = F\left(\mathbf{x}, t - \frac{R}{c}\right); \tag{1.6.6}$$

$[F]$ is to be considered as a function of $\mathbf{x}$, $\mathbf{y}$ and t being held fixed. Then

$$\frac{\partial}{\partial x_\alpha}[F] = \left[\frac{\partial F}{\partial x_\alpha}\right] - \left[\frac{\partial F}{\partial t}\right]\frac{R_\alpha}{cR}. \tag{1.6.7}$$

Now let p be a solution of the wave equation (1.6.1) with continuous derivatives up to the second order. Then (1.6.7) applied twice in succession to p gives

$$\nabla^2[p] = [\nabla^2 p] - \frac{2\mathbf{R}}{cR}\cdot\left[\operatorname{grad}\frac{\partial p}{\partial t}\right] + \frac{1}{c^2}\left[\frac{\partial^2 p}{\partial t^2}\right] - \frac{2}{cR}\left[\frac{\partial p}{\partial t}\right].$$

Also

$$\operatorname{grad}\left[\frac{\partial p}{\partial t}\right] = \left[\operatorname{grad}\frac{\partial p}{\partial t}\right] - \frac{\mathbf{R}}{cR}\left[\frac{\partial p}{\partial t}\right].$$

Eliminating $[\operatorname{grad} \partial p/\partial t]$ between these two equations one obtains

$$\nabla^2[p] + \frac{2\mathbf{R}}{cR}\cdot\operatorname{grad}\left[\frac{\partial p}{\partial t}\right] + \frac{2}{cR}\left[\frac{\partial p}{\partial t}\right] = \left[\nabla^2 p - \frac{1}{c^2}\frac{\partial^2 p}{\partial t^2}\right] = 0.\ddagger \tag{1.6.8}$$

† See, for instance, Lamb (1932), p. 494, or Courant-Hilbert (1937), p. 159.

‡ This equation is a particular case of the general intrinsic relation between p and $\partial p/\partial t$ that holds on a characteristic, (3.7.1).

After multiplication by $1/R$ this equation can be written as

$$\operatorname{div}\left\{\frac{1}{R}\operatorname{grad}[p]+\frac{\mathbf{R}}{R^3}[p]+\frac{2\mathbf{R}}{cR^2}\left[\frac{\partial p}{\partial t}\right]\right\}=0,$$

or, by (1.6.7),

$$\operatorname{div}\left\{\frac{1}{R}[\operatorname{grad}p]+\frac{\mathbf{R}}{R^3}[p]+\frac{\mathbf{R}}{cR^2}\left[\frac{\partial p}{\partial t}\right]\right\}=0. \qquad (1.6.9)$$

Let us integrate this over the domain $\epsilon \leqslant R \leqslant ct$, where $\epsilon > 0$, and transform the volume integral into a surface integral by the divergence theorem. The contribution from the surface $R=\epsilon$ is

$$-4\pi p(\mathbf{y}, t)+O(\epsilon).$$

Hence we obtain by making $\epsilon \to 0$ that

$$4\pi p(\mathbf{y}, t)=\int_{R=ct}\left\{\frac{1}{R}[\operatorname{grad} p]+\frac{\mathbf{R}}{R^2}[p]+\frac{\mathbf{R}}{cR^2}\left[\frac{\partial p}{\partial t}\right]\right\}\cdot\frac{\mathbf{R}}{R}R^2\,\mathrm{d}\Omega, \qquad (1.6.10)$$

where $\mathrm{d}\Omega$ is the element of area on the unit sphere. But when $R=ct$, then by (1.6.6)

$$[p]=p_0(\mathbf{x}),$$

$$\left[\frac{\partial p}{\partial t}\right]=p_1(\mathbf{x}),$$

$$[\operatorname{grad}p]=\operatorname{grad}p_0(\mathbf{x}).$$

Hence

$$4\pi p(\mathbf{y}, t)=t\int p_1(\mathbf{y}+\mathbf{R})\,\mathrm{d}\Omega+\int p_0(\mathbf{y}+\mathbf{R})\,\mathrm{d}\Omega$$
$$+ct\int\left\{\frac{\mathbf{R}}{R}\cdot\operatorname{grad}p_0(\mathbf{y}+\mathbf{R})\right\}\mathrm{d}\Omega,$$

with $|\mathbf{R}|=ct$, and this is obviously equivalent to (1.6.3).†

The most important feature of Poisson's solution (1.6.3) is that the pressure at a point $\mathbf{x}$ at time t depends only on the initial state of the medium at those points which are on the surface of a sphere of radius ct with centre $\mathbf{x}$. If the disturbance is initially confined to a domain D, then at a point $\mathbf{x}$ exterior to D the medium remains

† It should be noted that what has actually been proved is that if the initial-value problem (1.6.2) has a twice continuously differentiable solution, then that solution is necessarily (1.6.3). To make the argument conclusive one would have to verify that Poisson's solution satisfies the wave equation, the initial conditions, and the regularity hypothesis. This point is dealt with in detail in Hadamard (1923). See also the appendix to chapter 5.

at rest until ct is equal to the least distance of $\mathbf{x}$ from D. Thus the arrival time of the disturbance is given by Fermat's principle of minimum travel time, and the front of the disturbance is a wave front in the sense of geometrical optics. We may refer to this important result as the *propagation property* of the wave equation. The main object of the next chapter will be to establish an analogous property in the more general case of an inhomogeneous medium, and to discuss its implications for boundary-value problems; we shall then obviously not be able to argue from an explicit solution such as (1.6.3).

If the initially disturbed domain D is bounded, then the disturbance at any point $\mathbf{x}$ ends when ct is equal to the least upper bound of the distances $|\mathbf{x}-\mathbf{x}'|$, where $\mathbf{x}'$ may be any point of D. This is a particular property of the wave equation in a homogeneous medium, and does not hold in general. Even in a homogeneous medium it is not valid when reflectors are present.

CHAPTER 2

WAVE FRONTS AND CHARACTERISTICS

1. Introduction

A disturbance represented by a solution of the wave equation has the characteristic property that if it is confined at one instant, say $t=0$, to a domain D_0, then the domain D_t affected by it at a subsequent time t is deducible from D_0 by geometrical optics. In fact, the boundary of D_t is a wave front advancing into the undisturbed medium. We have already proved this for a homogeneous unlimited medium by means of Poisson's solution of the initial-value problem. The main object of this chapter is to establish this property in the general case, and to extend it to diffraction and reflexion. The propagation property is a consequence of another, and more fundamental, property of the wave equation. The state of the medium (represented, say, by the values of p and of $\partial p/\partial t$) at a point $\mathbf{x}$ at time t depends on the state at a previous time t_0 only at those points $(\mathbf{x}', t_0)$ which are such that a disturbance starting at a point $\mathbf{x}'$ and propagated away from it with the speed of sound can reach $\mathbf{x}$ in a time not exceeding $t-t_0$. This property can be expressed concisely in terms of space-time by saying that each point of space-time has a dependence domain, and that any disturbance localized at a point of space-time can only affect a domain called the influence domain of that point.

The method by which these basic properties will be proved is based on a simple integral identity which can be interpreted as the theorem of conservation of energy applied to a portion of the medium with a moving boundary. It is due to Zaremba (1915) and was subsequently rediscovered by Rubinowicz (1920) and by Friedrichs and Lewy (1928).† Diffraction was considered by the author (Friedlander 1949). The main argument is given in §§4 and 5; it is applied to reflexion problems in §6. The first two sections are preparatory. They deal respectively with some aspects of the geometry of space-time, and with the elements of geometrical

† An account of this method can be found in Courant-Hilbert (1937).

optics. Some further details concerning geometrical optics and the geometry of characteristics are given in the appendix to the chapter.

2. Space-time

The physical quantities which characterize the state of the medium are functions of position and of the time. One may think of them as specifying a succession of states in space which depend on the time as a parameter; this may be called the description in terms of space and time. Alternatively, one can think of x_1, x_2, x_3 and t as the coordinates of a point $(\mathbf{x}, t)$ in a four-dimensional continuum, and the physical quantities as functions defined over it; this is the description in space-time. The space-time description is particularly useful in the mathematical analysis, but that in terms of space and time more natural in physical applications. We begin by noting some simple points which are frequently used.

We shall have to deal with manifolds of three, two and one dimension in space-time. These will be called 3-spaces, 2-spaces and 1-spaces or curves respectively. Domains of space-time will be denoted by Greek capitals $\Sigma, \Omega, \ldots$; domains of space and 3-spaces in space-time by italic capitals $R, S, \ldots$; surfaces in space and 2-spaces in space-time by cursive capitals $\mathscr{R}, \mathscr{S}, \ldots$. The corresponding elements of integration will be denoted by $\mathrm{d}\Sigma$, $\mathrm{d}S$ and $\mathrm{d}\mathscr{S}$ respectively.

A 3-space S whose equation is

$$F(\mathbf{x}, t) = 0 \tag{2.2.1}$$

will be said to be smooth if either F or some appropriate function of F has continuous partial derivatives of the first order which do not vanish simultaneously on S. For such a 3-space we define the components (n_1, n_2, n_3, n_0) of the unit normal by the equations

$$\frac{n_1}{\partial F/\partial x_1} = \frac{n_2}{\partial F/\partial x_2} = \frac{n_3}{\partial F/\partial x_3} = \frac{n_0}{\partial F/\partial t}, \quad n_\alpha n_\alpha + n_0^2 = 1, \tag{2.2.2}$$

and the element of integration $\mathrm{d}S$ by

$$\pm n_0 \,\mathrm{d}S = \mathrm{d}x_1 \mathrm{d}x_2 \mathrm{d}x_3 = \mathrm{d}x, \quad \pm n_1 \,\mathrm{d}S = \mathrm{d}x_2 \,\mathrm{d}x_3 \,\mathrm{d}t, \quad \ldots, \tag{2.2.3}$$

where an appropriate sense of the normal must be chosen, as indicated by the ambiguity of sign.

The projections onto space of the intersections of the 3-planes $t = \text{constant}$ with S form a family of space surfaces depending on t as a parameter; thus S represents a surface $\mathscr{S}_t$ moving in space. A curve $\{\mathbf{x}(s), t(s)\}$ on S therefore represents a point $\mathbf{x}(t)$ in space which lies, at time t, on $\mathscr{S}_t$. If we denote the velocity of this point by V then we have, since $F\{x(t), t\} = 0$, by (2.2.2),

$$n_\alpha \dot{x}_\alpha + n_0 = 0, \quad \dot{x}_\alpha \dot{x}_\alpha = V^2 \quad (\dot{x}_\alpha = \mathrm{d}x_\alpha/\mathrm{d}t). \tag{2.2.4}$$

This implies at once that

$$V^2 \geqslant \frac{n_0^2}{n_\alpha n_\alpha},$$

with equality only when the $\dot{x}_\alpha$ are proportional to the n_α. Now the n_α are direction ratios of the normal to $\mathscr{S}_t$, so that in the case of equality V becomes the velocity v of $\mathscr{S}_t$ normal to itself at any point. We have therefore

$$v^2 = \frac{n_0^2}{n_\alpha n_\alpha}. \tag{2.2.5}$$

Furthermore, if ν_α are the direction cosines of the unit normal to $\mathscr{S}_t$ in the direction of propagation, then

$$n_0 \nu_\alpha + n_\alpha v = 0. \tag{2.2.6}$$

We shall often assume that (2.2.1) can be replaced by an equation of the form

$$t = \sigma(\mathbf{x}), \tag{2.2.7}$$

or that a 3-space can be divided into a finite number of portions each of which has an equation of this form. One then has, by (2.2.2) and (2.2.5),

$$\frac{\partial \sigma}{\partial x_\alpha} = -\frac{n_\alpha}{n_0} \quad (\alpha = 1, 2, 3), \tag{2.2.8}$$

$$v = \left(\frac{\partial \sigma}{\partial x_\alpha}\frac{\partial \sigma}{\partial x_\alpha}\right)^{-\frac{1}{2}}. \tag{2.2.9}$$

3. Characteristics and geometrical optics

A 3-space for which

$$n_\alpha n_\alpha = n_0^2 c^2 \tag{2.3.1}$$

is called a characteristic of the wave equation. The importance of the characteristics in the theory of the wave equation will be made clear in the next two sections and in chapter 3. But as a preliminary we must consider the geometry of characteristics briefly. By

(2.2.5), a characteristic represents a surface in space which propagates normal to itself with the speed of sound. Such a surface is a wave front in the sense of geometrical optics, and consequently the theory of the characteristics is equivalent to geometrical optics.

Let us assume the equation of a characteristic in the form

$$t=\tau(\mathbf{x}). \tag{2.3.2}$$

Then by (2.2.9) $\tau(\mathbf{x})$ satisfies a partial differential equation of the first order,

$$|\operatorname{grad}\tau|^2=\frac{1}{c^2}. \tag{2.3.3}$$

This is the *eikonal equation*. The rays are the curves in space which are orthogonal to the wave fronts (2.3.2); hence they are given by

$$\frac{d\mathbf{x}}{d\nu}=\operatorname{grad}\tau, \tag{2.3.4}$$

where ν is a parameter. The curves in space-time on the characteristic (2.3.2) whose projections onto space are the rays are the characteristics of the eikonal equation; following Hadamard, we shall call them the bicharacteristics of the wave equation. They satisfy (2.3.4) and

$$\frac{dt}{d\nu}=\frac{1}{c^2}, \tag{2.3.5}$$

which is a consequence of (2.3.2), (2.3.4) and (2.3.3). If τ is given then a unique solution of the differential equations (2.3.4) and (2.3.5) is determined by the values of $\mathbf{x}$, say $\mathbf{x}_0$, and of t, say t_0, for $\nu=0$. Since also $t_0=\tau(\mathbf{x}_0)$, it follows that a given characteristic contains a two-parameter family of bicharacteristics.† But we can, in fact, construct all characteristics as the loci of the points on suitably chosen two-parameter families of bicharacteristics.

It follows from (2.3.4) that

$$\frac{d}{d\nu}\left(\frac{\partial\tau}{\partial x_\alpha}\right)=\frac{\partial^2\tau}{\partial x_\alpha\,\partial x_\beta}\frac{dx_\beta}{d\nu}=\frac{\partial^2\tau}{\partial x_\alpha\,\partial x_\beta}\frac{\partial\tau}{\partial x_\beta}=\frac{1}{2}\frac{\partial}{\partial x_\alpha}|\operatorname{grad}\tau|^2,$$

whence by (2.3.3)

$$\frac{d}{d\nu}\left(\frac{\partial\tau}{\partial x_\alpha}\right)=\frac{1}{2}\frac{\partial}{\partial x_\alpha}\left(\frac{1}{c^2}\right). \tag{2.3.6}$$

† It must be assumed that the $\partial\tau/\partial x_\alpha$ and $1/c^2$ satisfy Lipschitz conditions.

Hence putting $\operatorname{grad}\tau = \mathbf{y}$ we obtain a system of seven ordinary differential equations for the unknowns $\mathbf{x}$, t, $\mathbf{y}$:

$$\frac{d\mathbf{x}}{d\nu} = \mathbf{y}, \quad \frac{d\mathbf{y}}{d\nu} = \tfrac{1}{2}\operatorname{grad}\left(\frac{1}{c^2}\right), \quad \frac{dt}{d\nu} = \frac{1}{c^2}. \tag{2.3.7}$$

A solution of this system is uniquely determined when $\mathbf{x}(0) = \mathbf{x}_0$, $t(0) = t_0$ and $\mathbf{y}(0) = \mathbf{y}_0$ are given, provided that the $(\partial/\partial x_\alpha)(1/c^2)$ satisfy Lipschitz conditions. Such a solution is, however, not in general a bicharacteristic. It follows from (2.3.7) that

$$\frac{d}{d\nu}\left(y^2 - \frac{1}{c^2}\right) = 0, \tag{2.3.8}$$

and hence a solution of (2.3.7) will be a bicharacteristic only when the initial values are restricted by the condition

$$y_0^2 = \frac{1}{c_0^2}, \quad c_0 = c(\mathbf{x}_0). \tag{2.3.9}$$

The quantities $(\mathbf{x}, t, \mathbf{y})$ determine a surface element in space-time, that is to say a point $(\mathbf{x}, t)$ and a plane through this point normal to the 4-vector $(y_1, y_2, y_3, -1)$. A solution of (2.3.7) is a one-parameter family of surface elements (a strip) and our analysis implies that a characteristic which contains a given surface element contains the corresponding strip. Two characteristics which touch at a point therefore touch along a common bicharacteristic through that point. Hence an envelope of characteristics is necessarily another characteristic.

Now consider a two-parameter family of bicharacteristics. It is given by a set of solutions of (2.3.7), $\mathbf{x}(\lambda, \mu, \nu)$, $t(\lambda, \mu, \nu)$, $\mathbf{y}(\lambda, \mu, \nu)$ which depends on two auxiliary parameters λ and μ. Let

$$\mathbf{x}(\lambda, \mu, 0) = \mathbf{x}_0(\lambda, \mu), \quad t(\lambda, \mu, 0) = t_0(\lambda, \mu), \quad \mathbf{y}(\lambda, \mu, 0) = \mathbf{y}_0(\lambda, \mu), \tag{2.3.10}$$

and suppose that (2.3.9) is satisfied by $\mathbf{x}_0$ and $\mathbf{y}_0$, that

$$dt_0 = \mathbf{y}_0 \cdot d\mathbf{x}_0, \tag{2.3.11}$$

and that

$$\left[\frac{D(x_1, x_2, x_3)}{D(\lambda, \mu, \nu)}\right]_{\nu=0} \neq 0. \tag{2.3.12}$$

This inequality implies that λ, μ and ν can be expressed as functions of $\mathbf{x}$ for sufficiently small values of $|\nu|$. Then it follows from

Cauchy's theory of partial differential equation of the first order† that

$$\tau(\mathbf{x}) = t\{\lambda(\mathbf{x}), \mu(\mathbf{x}), \nu(\mathbf{x})\} \tag{2.3.13}$$

satisfies the eikonal equation, so that $t = \tau(x)$ is a characteristic. Also

$$\operatorname{grad} \tau = \mathbf{y}\{\lambda(\mathbf{x}), \mu(\mathbf{x}), \nu(\mathbf{x})\}. \tag{2.3.14}$$

The equations (2.3.11) and (2.3.9) can be solved for $\mathbf{y}_0$ when $\mathbf{x}_0$ and t_0 are given as functions of λ and μ. We have here therefore a procedure for the determination of the characteristics that contain a given 2-space. The details of the calculation will not be developed here; they are discussed in the appendix to this chapter.

The equivalence of the eikonal equation and of Fermat's principle of minimum optical path can be established as follows. The metric of geometrical optics is defined by the line element

$$ds^2 = \frac{|\,d\mathbf{x}\,|^2}{c^2}. \tag{2.3.15}$$

The length of the optical path between two points $\mathbf{x}_0$ and $\mathbf{x}$ is the *geodesic distance*

$$\varpi(\mathbf{x}_0, \mathbf{x}) = \int_{\mathbf{x}_0}^{\mathbf{x}} \frac{|\,d\mathbf{x}\,|}{c} = \varpi(\mathbf{x}, \mathbf{x}_0), \tag{2.3.16}$$

the integral being taken along a geodesic. If $\mathbf{x} = \mathbf{x}(\sigma)$ is the equation of this geodesic in terms of a parameter and $\mathbf{x}(0) = \mathbf{x}_0$, then

$$d\varpi(\mathbf{x}_0, \mathbf{x}) = \frac{\mathbf{x}'(\sigma)}{|\,\mathbf{x}'(\sigma)\,|} \cdot \frac{d\mathbf{x}}{c} - \frac{\mathbf{x}'(0)}{|\,\mathbf{x}'(0)\,|} \cdot \frac{d\mathbf{x}_0}{c_0}, \tag{2.3.17}$$

according to the fundamental formula of the calculus of variations. Hence ϖ satisfies the eikonal equation both as a function of $\mathbf{x}$ and of $\mathbf{x}_0$.

Now let $\mathscr{S}_0$ be a given surface and $\mathbf{x}$ a point not on $\mathscr{S}_0$. Then the least geodesic distance of $\mathbf{x}$ from $\mathscr{S}_0$ is given by the minimum of the integral (2.3.16) when $\mathbf{x}$ is fixed and $\mathbf{x}_0$ describes $\mathscr{S}_0$. The integral must obviously be taken along a geodesic joining $\mathbf{x}_0$ and $\mathbf{x}$. By (2.3.17) it will therefore be stationary with respect to variations of $\mathbf{x}_0$ only if

$$\mathbf{x}_0' \,.\, d\mathbf{x}_0 = 0, \tag{2.3.18}$$

† See, for instance, Courant-Hilbert (1937), pp. 66–8.

where $d\mathbf{x}_0$ is a displacement tangential to $\mathscr{S}_0$. This condition determines $\mathbf{x}_0$ as a function of $\mathbf{x}$, and the least geodesic distance $\tau(\mathbf{x})$ of $\mathbf{x}$ from $\mathscr{S}$ is

$$\tau(\mathbf{x}) = \varpi(\mathbf{x}_0, \mathbf{x}).$$

When $\mathbf{x}$ is changed, $\mathbf{x}_0$ changes also and $d\tau$ is given by (2.3.17). But as the displacement of $\mathbf{x}_0$ is necessarily tangential to $\mathscr{S}_0$ it follows from (2.3.18) that

$$d\tau = d\mathbf{x} \,.\, \text{grad}\, \tau = \frac{\mathbf{x}'(0)}{|\,\mathbf{x}'(\sigma)\,|} \cdot \frac{d\mathbf{x}}{c}.$$

Hence $\tau(\mathbf{x})$ satisfies the eikonal equation. It follows that $\tau(\mathbf{x}) = t$ is the equation of a wave front which reduces to the given surface $\mathscr{S}_0$ for $t = 0$; the arrival time at the point $\mathbf{x}$ is the least geodesic distance of the point from $\mathscr{S}_0$. If there is more than one geodesic through $\mathbf{x}$ normal to $\mathscr{S}_0$ then the situation is complicated by the presence of caustics; these are discussed in the appendix to this chapter.

A case of particular importance is that where $\mathscr{S}_0$ reduces to a single point $\mathbf{x}_0$. The pseudo-spherical fronts diverging from this point from the instant t_0 are

$$\varpi(\mathbf{x}_0, \mathbf{x}) = t - t_0 \quad (t \geqslant t_0). \tag{2.3.19}$$

This equation represents a characteristic situated in $t \geqslant t_0$. Since obviously

$$\varpi(\mathbf{x}_0, \mathbf{x}) = \frac{|\,\mathbf{x} - \mathbf{x}_0\,|}{c_0} + o(|\,\mathbf{x} - \mathbf{x}_0\,|) \tag{2.3.20}$$

as $|\,\mathbf{x} - \mathbf{x}_0\,| \to 0$, this characteristic touches the semi-cone

$$t - t_0 = \frac{|\,\mathbf{x} - \mathbf{x}_0\,|}{c_0}$$

at $(\mathbf{x}_0, t_0)$. It is therefore called a *characteristic semi-conoid* with vertex $(\mathbf{x}_0, t_0)$. There is a second characteristic semi-conoid with this vertex whose equation is

$$\varpi(\mathbf{x}_0, \mathbf{x}) = t_0 - t \quad (t \leqslant t_0). \tag{2.3.21}$$

It is situated in $t \leqslant t_0$ and represents wave fronts converging toward $\mathbf{x}_0$. The two semi-conoids together make up the *characteristic conoid* with vertex $(\mathbf{x}_0, t_0)$,

$$(t - t_0)^2 - [\varpi(\mathbf{x}_0, \mathbf{x})]^2 = 0. \tag{2.3.22}$$

The principle that an envelope of a family of characteristics is again a characteristic now furnishes an alternative method for the construction of the characteristics that contain a given 2-space. They are simply the envelopes of the characteristic conoids whose vertices are the points of the 2-space. This is in effect Monge's method of integrating a partial differential equation of the first order.

The eikonal equation, the equations of the bicharacteristics, the definition of the geodesic distance ϖ and of the characteristic semi-conoids are sufficient for most of the applications of the geometry of the characteristics and of geometrical optics which will be made in this book. But for the sake of completeness some further details concerning the geometry of the characteristics have been included, which will be found in the appendix to this chapter.

4. The uniqueness theorem; dependence and influence domains

Turning now to the main subject of this chapter, let us consider a solution $\psi(\mathbf{x}, t)$ of the homogeneous wave equation (1.3.1) with continuous derivatives up to the second order at least.† Thus

$$\frac{1}{\rho c^2}\frac{\partial^2\psi}{\partial t^2} - \operatorname{div}\left(\frac{1}{\rho}\operatorname{grad}\psi\right) = 0.$$

Now

$$0 = 2\frac{\partial\psi}{\partial t}\left\{\frac{1}{\rho c^2}\frac{\partial^2\psi}{\partial t^2} - \operatorname{div}\left(\frac{1}{\rho}\operatorname{grad}\psi\right)\right\}$$

$$= \frac{\partial}{\partial t}\left\{\frac{1}{\rho c^2}\left(\frac{\partial\psi}{\partial t}\right)^2\right\} - 2\operatorname{div}\left(\frac{1}{\rho}\frac{\partial\psi}{\partial t}\operatorname{grad}\psi\right) + \frac{2}{\rho}\operatorname{grad}\psi\,.\operatorname{grad}\frac{\partial\psi}{\partial t}$$

or

$$\frac{\partial}{\partial t}\left\{\frac{1}{\rho c^2}\left(\frac{\partial\psi}{\partial t}\right)^2 + \frac{1}{\rho}|\operatorname{grad}\psi|^2\right\} - \operatorname{div}\frac{1}{\rho}\operatorname{grad}\frac{\partial\psi}{\partial t} = 0. \qquad (2.4.1)$$

If ψ is the velocity potential defined by (1.3.5) then this equation is equivalent to

$$\frac{\partial}{\partial t}\left(\tfrac{1}{2}\rho u^2 + \frac{p^2}{2\rho c^2}\right) + \operatorname{div}(p\mathbf{u}) = 0.$$

† This regularity hypothesis will be found convenient in connexion with the developments of chapter 3. It could however be lightened; it is, for instance, sufficient to assume that the first-order derivatives are continuous and that the second-order derivatives are piece-wise continuous and integrable.

Adding the equation (1.3.9) multiplied by the pressure p_0 of the undisturbed medium and integrating over a bounded domain of space S with frontier $\mathscr{S}$ we obtain

$$\frac{\partial}{\partial t}\int_S \left\{\tfrac{1}{2}\rho u^2 + \frac{(2p_0+p)p}{2\rho c^2}\right\} \mathrm{d}x = \int_{\mathscr{S}} (p_0+p)\, u_n \, \mathrm{d}\mathscr{S},$$

where u_n is the component of particle velocity along the normal to $\mathscr{S}$ drawn into S. This equation is usually interpreted as the energy equation; the right-hand side is the rate at which the fluid outside S does work on that inside S and the left-hand side therefore represents the rate of change of the energy of the fluid inside S. The term $\frac{1}{2}\rho u^2$ is clearly the kinetic energy per unit volume, and $\{(2p_0+p)p\}/2\rho c^2$ must be thought of as the potential energy per unit volume.†

The physical interpretation of (2.4.1) is, however, irrelevant to our purpose. We are concerned with its analytic consequences, and these can only be brought out by integrating it over a general bounded domain Σ of space-time with piece-wise smooth frontier S. The four-dimensional divergence theorem then gives

$$\int_S \left\{ n_0 \left[\frac{1}{c^2}\left(\frac{\partial \psi}{\partial t}\right)^2 + \frac{\partial \psi}{\partial x_\alpha}\frac{\partial \psi}{\partial x_\alpha}\right] - 2\frac{\partial \psi}{\partial t}\left(n_\alpha \frac{\partial \psi}{\partial x_\alpha}\right)\right\}\frac{\mathrm{d}S}{\rho} = 0, \qquad (2.4.2)$$

where (n_1, n_2, n_3, n_0) are the direction cosines of the unit normal to S drawn away from Σ. On any part of S on which n_0 does not change sign, the integrand can be put into the form

$$\left\{\sum_{\alpha=1}^{3}\left(\frac{\partial \psi}{\partial x_\alpha} - \frac{n_\alpha}{n_0}\frac{\partial \psi}{\partial t}\right)^2 + \left(\frac{1}{c^2} - \frac{n_\alpha n_\alpha}{n_0^2}\right)\left(\frac{\partial \psi}{\partial t}\right)^2\right\}\frac{n_0}{\rho}. \qquad (2.4.3)$$

It is therefore a definite quadratic form in $\partial\psi/\partial t$ and the $\partial\psi/\partial x_\alpha$ on a space-like part of S, and a semi-definite quadratic form on any characteristic part of S.

Suppose now that ψ and $\partial\psi/\partial t$ are given for $t=0$; this is the initial-value problem. The data determine the solution (whose

† The energy equation contains terms of the second order, but is deduced from the linearized equations of motion. It is therefore not surprising that it differs from the equation obtained from the exact energy equation by retaining terms of the first and second order. It must be recognized that the identification of the term $\{(2p_0+p)\,p\}/2\rho c^2$ with the potential energy of deformation of the medium is a convention, and not necessarily an approximation of the second order to the intrinsic energy, except when the medium is uniform.

existence we assume) in both $t>0$ and $t<0$; since the wave equation remains unchanged when t is replaced by $-t$, it is sufficient to consider one of these cases only, say $t>0$. Let $(\mathbf{x}_0, t_0)$ be any point of space-time, with $t_0>0$. Then the characteristic half-conoid $C: t=t_0-\varpi(\mathbf{x}_0, \mathbf{x})$, the 3-plane $t=0$, and a 3-plane $t=t'$, where $0 \leqslant t' < t_0$, bound a domain Σ to which (1.3.2) can be applied (fig. 2.1). On $t=0$, one has $n_0=-1$, $n_1=n_2=n_3=0$; on $t=t'$, one

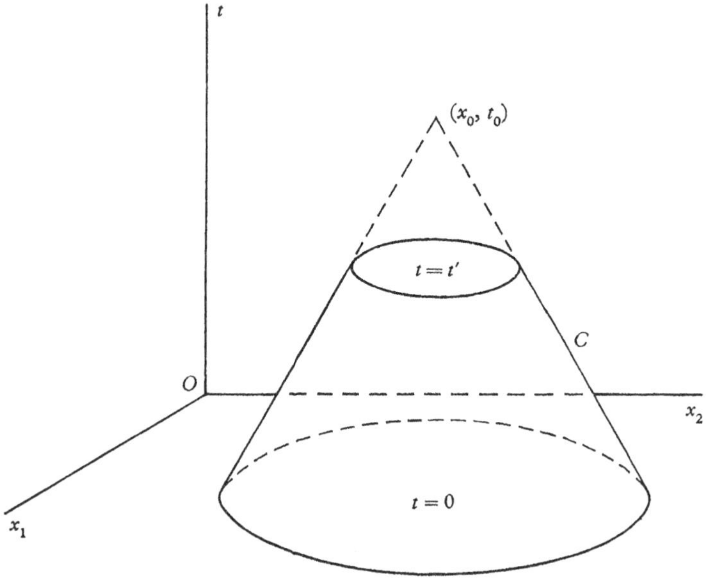

Fig. 2.1. Domain of integration for the proof of the uniqueness theorem.

has $n_0=1$, $n_1=n_2=n_3=0$; C is a characteristic, and $n_0>0$ on C. Hence the contribution from C is non-negative, and has the same sign as that from $t=t'$; it therefore follows from (2.4.2) that

$$\int_{\varpi \leqslant t_0 - t'} \left[\frac{1}{c^2}\left(\frac{\partial \psi}{\partial t}\right)^2 + \frac{\partial \psi}{\partial x_\alpha}\frac{\partial \psi}{\partial x_\alpha}\right]_{t=t'} \frac{\mathrm{d}x}{\rho}$$

$$\leqslant \int_{\varpi \leqslant t_0} \left[\frac{1}{c^2}\left(\frac{\partial \psi}{\partial t}\right)^2 + \frac{\partial \psi}{\partial x_\alpha}\frac{\partial \psi}{\partial x_\alpha}\right]_{t=0} \frac{\mathrm{d}x}{\rho}. \quad (2.4.4)$$

This inequality expresses the fact that in a certain sense the solution depends continuously on the initial values. It is an instance of a class of inequalities valid for the solutions of the wave equation and their derivatives which play an important part in abstract (nonconstructive) existence proofs. We shall not pursue this line of argument here, but restrict ourselves to a very simple deduction. Suppose in fact that, for $t=0$, $\psi=\partial\psi/\partial t=0$ when $\varpi(\mathbf{x}_0,\mathbf{x})\leqslant t_0$. Then it follows at once from (2.4.4) and the assumed continuity of the derivatives of ψ that

$$\frac{\partial\psi}{\partial t}=\frac{\partial\psi}{\partial x_1}=\frac{\partial\psi}{\partial x_2}=\frac{\partial\psi}{\partial x_3}=0 \quad (t=t',\ \varpi(\mathbf{x}_0,\mathbf{x})\leqslant t_0-t').$$

Since t' can be any number between 0 and t_0, ψ is therefore constant in $0\leqslant t\leqslant t_0-\varpi$. By continuity $\psi=0$ in $0\leqslant t\leqslant t_0-\varpi(\mathbf{x}_0,\mathbf{x})$, and in particular $\psi(\mathbf{x}_0,t_0)=0$. If the initial data determining two solutions of the wave equation coincide in $\varpi(\mathbf{x}_0,\mathbf{x})\leqslant t_0$, then these solutions are equal to each other in the whole of $0\leqslant t\leqslant t_0-\varpi(\mathbf{x}_0,\mathbf{x})$. But the choice of the origin of time is arbitrary; we can therefore formulate this result a little differently. Let us call the 'interior' of the characteristic half-conoid with vertex $(\mathbf{x}_0,t_0)$ situated in $t\leqslant t_0$ the *dependence domain* of $(\mathbf{x}_0,t_0)$, $\Delta(\mathbf{x}_0,t_0)$. The points $(\mathbf{x},t)$ of Δ satisfy the inequality

$$t\leqslant t_0-\varpi(\mathbf{x}_0,\mathbf{x}). \tag{2.4.5}$$

We can then say that $\psi(\mathbf{x}_0,t_0)$ *is determined by initial data at an earlier time t if these data are given at all points* (x,t_1) *which are in the dependence domain of* $(\mathbf{x}_0,t_0)$. These points constitute the geodesic sphere $\varpi(\mathbf{x}_0,\mathbf{x})\leqslant t_0-t_1$.

Conversely, we can consider the effect of a change in the data on the initial 3-plane $t=t_1$ on the solution at $(\mathbf{x}_0,t_0)$. Let us go, for a moment, beyond the continuity hypotheses made at the beginning of this section, and imagine that the flow is modified by a localized impulse applied at $\mathbf{x}_1$ at time t_1. Since we may think of this as the limit of a continuously distributed disturbance centred on $\mathbf{x}_1$, we see at once that $\psi(\mathbf{x},t)$ will be affected only if $(\mathbf{x}_1,t_1)\in\Delta(\mathbf{x},t)$. The set of all points $(\mathbf{x},t)$ which satisfy this will be called the *influence domain* of $(\mathbf{x}_1,t_1)$ (see fig. 2.2). Now the geodesic distance $\varpi(\mathbf{x}_1,\mathbf{x})$ is a symmetric function of the two points $\mathbf{x}_1$, $\mathbf{x}$; hence the influence domain $\Delta'(\mathbf{x}_1,t_1)$ is defined by

$$t\geqslant t_1+\varpi(\mathbf{x}_1,x) \tag{2.4.6}$$

and is the 'interior' of the characteristic half-conoid with vertex $(\mathbf{x}_1,t_1)$ situated in $t\geqslant t_1$. Obviously, a point $(\mathbf{x},t)$ is in the dependence domain of $(\mathbf{x}_0, t_0)$ if and only if $(\mathbf{x}_0, t_0)$ is in the influence domain of $(\mathbf{x}, t)$, and vice versa. The frontier of $\Delta'(\mathbf{x}_1, t_1)$ represents a wave front diverging from $\mathbf{x}_1$, starting at time t_1; the uniqueness theorem therefore expresses the simple physical fact that *changes in the field are propagated with the speed of sound.*

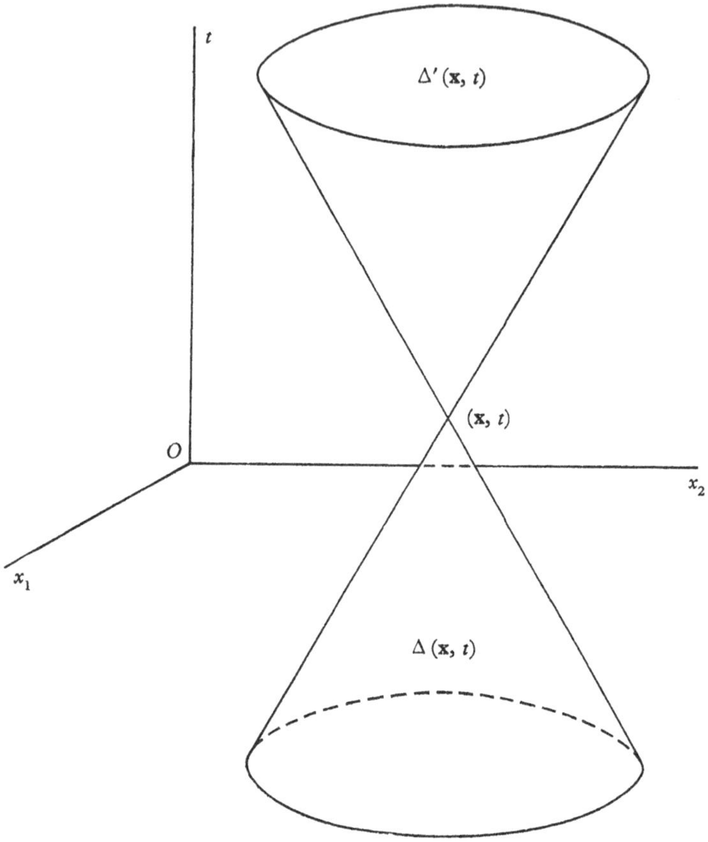

Fig. 2.2. Dependence and influence domains of a point in space-time.

Returning to the initial-value problem let us suppose that ψ and $\partial\psi/\partial t$ differ from zero in a domain B' of space, and are zero in the complementary domain B. If $\mathbf{x}$ is a point of B, then $\psi(\mathbf{x}, t) = 0$ as long as t does not exceed the least geodesic distance of x from B', defined by

$$\tau(\mathbf{x}) = \inf \varpi(\mathbf{x}, \mathbf{x}'), \quad \mathbf{x}' \in B'. \tag{2.4.7}$$

Hence the *arrival time of the disturbance at any point of B is given by Fermat's principle.* The 3-space S whose equation is $\tau(\mathbf{x}) = t$ is a characteristic, which divides $t \geqslant 0$ into an undisturbed and a disturbed domain. A point $(\mathbf{x}, t)$ will be in the disturbed domain if its dependence domain intercepts, on $t = 0$, points of B'. This implies that $(\mathbf{x}, t)$ is in the influence domain of at least one point $(\mathbf{x}', 0)$, where $\mathbf{x}' \in B$. Hence the disturbed domain is also the union of the influence domains of the points of B'; the wave front, which is the

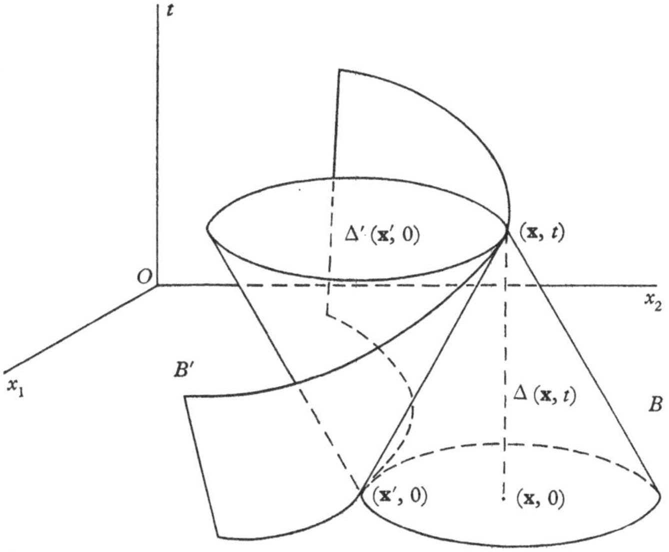

Fig. 2.3. Constructions of the characteristic representing the fronts of a disturbance.

projection onto space of the intersection of the frontier of the disturbed domain and the 3-planes $t = \text{const.}$, is therefore that envelope of the geodesic spheres with centres on the frontier of B' and radius t which is in the initially undisturbed region of space, B. This is *Huygens' construction* of the wave fronts (fig. 2.3).

We have tacitly assumed that a characteristic conoid has no singularities other than its vertex. In the most familiar example, that of a homogeneous medium, c is constant, $\varpi = |\mathbf{x}_0 - \mathbf{x}|/c$, and this assumption is valid. But in general it may happen that the rays diverging from a point have a caustic; the argument must then be modified. It can be shown that if c and its derivatives are bounded, and satisfy Lipschitz conditions uniformly, and c is bounded away from zero everywhere, then there exists a constant k such that, if $|\mathbf{x}_0 - \mathbf{x}| < k$, $\mathbf{x}_0$ and $\mathbf{x}$ can be joined by a unique geodesic. There is, then, no caustic in $|\mathbf{x}_0 - \mathbf{x}| < k$. The radius of the greatest geodesic sphere inscribed in $|\mathbf{x}_0 - \mathbf{x}| < k$ also has a lower bound T; hence the uniqueness theorem for the initial-value problem with data on $t = 0$ hold provided that $t \leqslant T$. But we can now consider the solution for $t \geqslant T$ to be determined by the supposedly known values of ψ and $\partial\psi/\partial t$ on $t = T$; these then determine ψ in $T \leqslant t \leqslant 2T$. Continuing in this way, we can define the solution of the initial-value problem uniquely for all $t \geqslant 0$. The definition of the dependence and influence domains of a point can be extended similarly. This extension conforms to the physical principle of determinacy, and to the principle which Hadamard has called Huygens' major premise. This asserts that, if $\psi(x, t)$ is calculated on the one hand from data on $t = 0$, and on the other hand from data at an intermediate time t' which have themselves been determined from the initial values, then the same value of ψ is obtained. In the absence of caustics, this principle is an obvious consequence of the uniqueness theorem.

5. Diffraction

The arguments of the preceding section apply only in an unlimited medium. But they can be extended to the case where the medium occupies a domain S of space ('physical space') bounded by a surface $\mathscr{R}$ on which a suitable boundary condition is prescribed. The boundary $\mathscr{R}$ need not be connected. We shall here assume that the boundary condition is (1.5.2), that is to say, in effect,

$$\frac{\partial\psi}{\partial\nu} + \lambda\frac{\partial\psi}{\partial t} = 0, \tag{2.5.1}$$

where $\partial/\partial\nu$ denotes differentiation along the normal to $\mathscr{R}$ drawn away from S, and $\lambda \geqslant 0$.

Let us again assume that ψ and $\partial\psi/\partial t$ are given for $t = 0$; these data can of course only be given in S. Let R denote the 3-cylinder parallel to the t-axis whose base is $\mathscr{R}$, that is to say, the set of points $(\mathbf{x}, t)$ with $\mathbf{x} \in \mathscr{R}$. If $\mathbf{x}_0$ is any point of S and $t_0 > 0$, then the intersection of $\Delta(\mathbf{x}_0, t_0)$ and of $t=0$ may be contained in S; in that case the argument of the preceding section applies. When this is not the case, we can apply the basic integral identity (2.4.2) to the domain bounded by $\Delta(\mathbf{x}_0, t_0)$, by $t=0$, by a 3-plane $t=t'$, where $0<t'<t_0$, and by R. As before, the contribution of Δ is non-negative, that of $t=t'$ is positive, that of $t=0$ negative. The integral over R is

$$-2\int \frac{\partial\psi}{\partial t} n_\alpha \frac{\partial\psi}{\partial x_\alpha}\frac{\mathrm{d}S}{\rho} = 2\int \lambda \left(\frac{\partial\psi}{\partial t}\right)^2 \frac{\mathrm{d}R}{\rho},$$

since the normal to R has the direction cosines $(\nu_1, \nu_2, \nu_3, 0)$, and as by hypothesis $\lambda \geqslant 0$ it is also non-negative. It follows that the fundamental inequality (2.4.4) still holds, provided that the integrals are extended only over those parts of the geodesic spheres $\varpi(\mathbf{x}_0, \mathbf{x}) \leqslant t_0$ and $\varpi(\mathbf{x}_0, \mathbf{x}) \leqslant t_0 - t'$ which are in S. The uniqueness theorem therefore follows as before.†

This result seems to imply that the fronts of disturbances are still given by Fermat's principle. This is indeed the case, but only if Fermat's principle is suitably modified. Suppose, as in §4, that $p=0$, $\partial p/\partial t = 0$ for $t=0$ except in a domain B of space, with frontier $\mathscr{B}$. Then the arrival time of the front at any point not in B is the least geodesic distance $\tau(\mathbf{x})$ of $\mathbf{x}$ *from B with respect to curves in S only*. It may happen that there are points $\mathbf{x}$ of S which are such that no geodesic $(\mathbf{x}, \mathbf{x}_0)$ can be constructed which joins $\mathbf{x}$ to $\mathbf{x}_0$ of $\mathscr{B}$, is normal to $\mathscr{B}$ at $\mathbf{x}_0$, and lies in S (or there may be such geodesics, but the geodesic distance of $\mathbf{x}$ from $\mathscr{B}$ along these is not an absolute minimum). When this is the case, we shall speak of *diffraction*; the points $\mathbf{x}$ which have the property just stated constitute the *shadow* in the sense of geometrical optics, with respect to an incident wave front whose position for $t=t_0$ is $\mathscr{B}$.

Let $\mathbf{x}$ be a point of the shadow. It is reasonable to assume that the disturbance must reach $\mathbf{x}$ eventually; let $\tau(\mathbf{x})$ be its arrival time.

† It will be noted that, whereas the uniqueness theorem proved in §3 can be 'reversed' and used for $t<0$ as well as for $t>0$, by the simple device of applying it to $\psi(\mathbf{x}, -t)$, the same procedure in the present case leads to the result that the solution is uniquely determined in $t<0$ only if $\lambda \leqslant 0$. The reason for this is that the boundary condition (1.4.1) implies that energy is lost at the boundary, where the pressure of the fluid does work on the boundary. This is an irreversible process, and if one wants the solution to 'run backwards' one must change the sign of λ. In the case of a fixed rigid boundary, or of a free boundary, there is no need to change the boundary condition.

There is a ray λ through $\mathbf{x}$ which is a geodesic of the metric (2.3.15) and is normal to the wave front $\tau(\mathbf{x})=t$ at $\mathbf{x}$. At any earlier instant t', the front meets this ray at a point $\mathbf{x}'$ such that $\varpi(\mathbf{x}',\mathbf{x})=\tau(\mathbf{x})-t'$. As t' is decreased, $\mathbf{x}'$ moves along the ray away from $\mathbf{x}$. By hypothesis, the ray cannot meet $\mathscr{B}_0$ (for it would, by virtue of Fermat's principle, have to be normal to $\mathscr{B}_0$ at its intersection, that is to say, be one of the rays associated with $\mathscr{B}_0$, and by hypothesis $\mathbf{x}$ is in the shadow and does not lie on any such ray); hence it must meet $\mathscr{R}$, the boundary. Let $\mathbf{x}_0$ denote this intersection and $\tau(\mathbf{x}_0)$ the arrival time of the disturbance there. The unit vector $\mathbf{q}_0$ tangential to the ray in the sense of propagation of the front (which is $c\nabla\tau$) must point into S as any point $\mathbf{x}^*$ adjacent to $\mathbf{x}_0$ on the ray and corresponding to the time $\tau(\mathbf{x}_0)+\delta t$ is in S. There are now two possibilities. Either $\mathbf{q}_0$ is tangential to the boundary, or it makes a finite angle with the tangent plane to $\mathscr{R}$ at $\mathbf{x}_0$. We can show that this second alternative must be ruled out. The front $\tau(\mathbf{x})=t$ meets $\mathscr{R}$ in a (moving) curve σ_t, which is the front of the disturbance propagated over $\mathscr{R}$ itself. This curve divides any neighbourhood $\mathscr{U}$ of $\mathbf{x}_0$ on $\mathscr{R}$ into an undisturbed domain $\mathscr{U}'$ and a disturbed domain. Let $\mathbf{x}_1$ be a point of $\mathscr{U}'$, and consider the geodesic distance of $\mathbf{x}_1$ from the points of $\tau(\mathbf{x})=\tau(\mathbf{x}_0)$. If $\mathbf{x}_1$ is near to $\mathbf{x}_0$, and the points $\mathbf{x}$ considered are sufficiently near to $\mathbf{x}_1$, then both $\tau(\mathbf{x})=\tau(\mathbf{x}_0)$ and $\mathscr{R}$ can be replaced by their tangent planes, and the geodesics by straight lines. Clearly, if $\mathbf{q}_0$ makes a finite angle with the tangent plane to $\mathscr{R}$ at $\mathbf{x}_0$, then the least geodesic distance of $\mathbf{x}_1$ from the points of σ_t is to be measured along $\mathscr{R}$ on a line normal to σ_t (fig. 2.4). Let us denote this by δ; then by the uniqueness theorem

$$\tau(\mathbf{x}_1)-\tau(\mathbf{x}_0)\geqslant\delta.$$

If we make the distance $|\mathbf{x}_1-\mathbf{x}_0|$ tend to zero then $c_0\delta\,|\,\mathbf{x}_1-\mathbf{x}_0\,|\to 1$, where $c_0=c(\mathbf{x}_0)$. Also $|\,\tau(\mathbf{x}_0)-\tau(\mathbf{x}_1)\,|/|\,\mathbf{x}_1-\mathbf{x}_0\,|\to 1/V_R$, where V_R is the velocity of propagation of σ_t normal to itself at $\mathbf{x}_0$; hence we have in the limit

$$\frac{1}{V_R}\geqslant\frac{1}{c_0}\quad\text{or}\quad V_R\leqslant c_0.$$

But V_R is also the velocity of a moving point which remains on $\tau(\mathbf{x})=t$; hence by (2.2.5) and the fact that $\tau(\mathbf{x})=t$ is a characteristic, $V_R\geqslant c_0$. It follows that $V_R=c_0$, that is the say, σ_t propagates on $\mathscr{R}$

normal to itself with the velocity of sound. Since the only curves along which the velocity of a point of $\tau(\mathbf{x})=t$ is equal to that of sound are the rays, and the rays are normal to the wave fronts, it follows that $\tau(\mathbf{x})=t$ is normal to $\mathscr{R}$ and that $\mathbf{q}_0$ is tangential to $\mathscr{R}$. We have therefore shown that *in a shadow the disturbance propagates over the reflector surface $\mathscr{R}$ with the speed of sound, and that the rays associated with the wave fronts in the shadow are tangential to $\mathscr{R}$.* Such fronts will be called *diffracted fronts*.

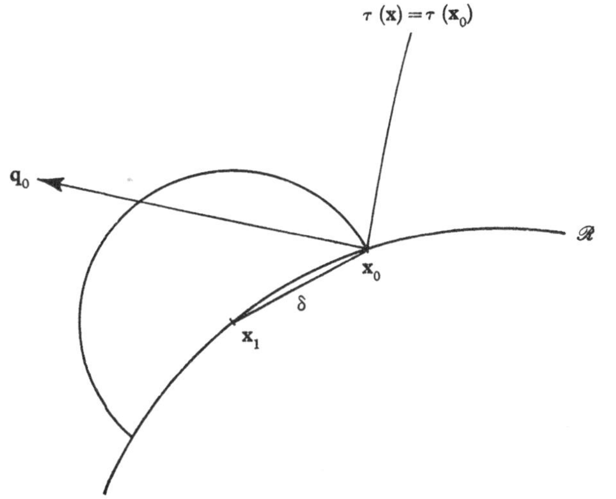

Fig. 2.4. Wave front advancing along a boundary with supersonic velocity.

It is now clear that the least geodesic distance of a point $\mathbf{x}$ of the shadow from the initial position $\mathscr{B}$ of the incident wave front is measured along arcs of geodesics which touch $\mathscr{R}$, and along curves of $\mathscr{R}$. This is in accordance with Fermat's principle, enunciated as a minimum travel time principle with respect to curves in S only.† There is a simple illustration in the case of a homogeneous medium ($c=$const.). Suppose that one end of a string can slide freely on $\mathscr{B}$ and that the string is tautened in such a manner that it passes through a given point $\mathbf{x}$ of S. Then the arrival time $\tau(\mathbf{x})$ of the disturbance at $\mathbf{x}$ is given by the distance of $\mathbf{x}$ from $\mathscr{B}$ measured along

† Hadamard (1910), pp. 183–8.

the string, divided by the velocity of sound (see fig. 2.5). A particularly simple case arises when the reflector is a half-plane or a wedge; this will be considered in chapter 5.

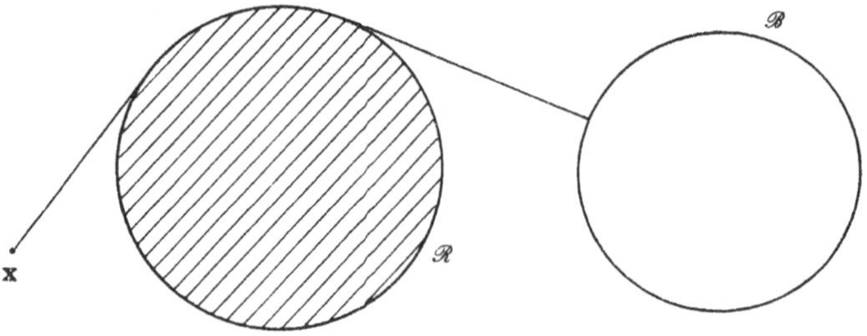

Fig. 2.5. The 'taut string' construction of diffracted rays.

6. Reflected fronts

Reflexion problems are usually formulated by giving an incident disturbance and asking for the secondary disturbance set up by the reflectors. Suppose that the incident disturbance is a sharp-fronted pulse, and that at some instant which we take to be $t=0$ its front $\mathscr{B}$ does not meet the boundary $\mathscr{R}$. With $\mathscr{B}$ we can associate the 'incident' rays, which are the geodesics normal to $\mathscr{B}$ starting at points of $\mathscr{B}$ and drawn into the initially undisturbed domain of S. We can now divide S into the shadow S'' which consists of all points of S that cannot be joined to $\mathscr{B}$ by an incident ray in S, and its complementary domain S'. In this the front of the disturbance is the wave front $\mathscr{B}_t$ determined by Fermat's principle as if the medium were unlimited; in S'' the front is a diffracted front whose rays are tangential to $\mathscr{R}$ and touch $\mathscr{R}$ at points in S''. Let the incident front $\mathscr{B}_t$ in S' have the equation $t=\tau_i(\mathbf{x})$; this is necessarily the equation of a characteristic B_i. The intersection of B_i and R is a 2-space $\mathscr{R}^*$ which represents the arrival of the incident front at $\mathscr{R}$. We can now construct a second characteristic B_r, with equation $t=\tau_r(x)$, which meets R in $\mathscr{R}^*$, as the envelope of the frontiers of influence domains whose vertices are points of $\mathscr{R}^*$, as far as this envelope lies in Σ, the space-time domain whose base is physical space. Then $\tau_r(x)=\tau_i(x)$ on $\mathscr{R}$.

By (2.3.3) and (2.3.4), we have

$$\operatorname{grad}\tau_i=\frac{1}{c}\mathbf{q}_i,\quad \operatorname{grad}\tau_r=\frac{1}{c}\mathbf{q}_r,\quad |\mathbf{q}_i|=|\mathbf{q}_r|=1, \tag{2.6.1}$$

where $\mathbf{q}_i$, $\mathbf{q}_r$ are unit tangent vectors alone the rays associated with the fronts represented by B_i, B_r respectively. Also, the normal $\boldsymbol{\nu}$ to $\mathscr{R}$ pointing into S is given by

$$\Lambda\boldsymbol{\nu} = \operatorname{grad}\tau_i - \operatorname{grad}\tau_r,$$

where $$\Lambda^2 = |\operatorname{grad}\tau_i - \operatorname{grad}\tau_r|^2 = \frac{2}{c^2}(1 - \mathbf{q}_i \cdot \mathbf{q}_r)$$

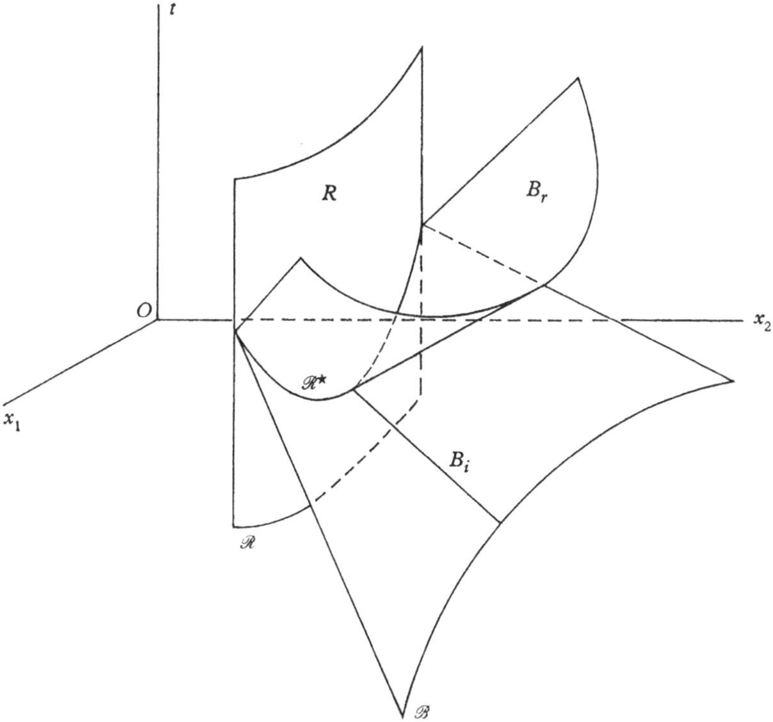

Fig. 2.6. Characteristics representing incident and reflected wave fronts.

by (2.6.1); let us put $\quad -\mathbf{q}_i \cdot \mathbf{q}_r = \cos 2\theta,$

so that 2θ is the angle between the two rays at $\mathscr{R}$. Then

$$2\boldsymbol{\nu}\cos\theta = \mathbf{q}_i - \mathbf{q}_r. \tag{2.6.2}$$

This equation implies the usual *reflexion law* of geometrical optics: $-\mathbf{q}_i$ and $\mathbf{q}_r$ *are coplanar with the normal* $\boldsymbol{\nu}$ *and equally inclined to it.* We naturally call the rays associated with B_r *reflected rays* (see fig. 2.6).

Since $\tau_r(x)$ is the time taken by a point travelling with velocity c from $\mathscr{B}$ to $\mathscr{R}$ along an incident ray and then to $\mathbf{x}$ along a reflected ray, it is clear that $\tau_r(\mathbf{x}) > \tau_i(\mathbf{x})$ except at $\mathscr{R}$ where there is equality. Let us now suppose the medium to be extended to the domain of space excluded by the reflector in any manner whatever, and consider the disturbance which has, for $t=0$, the same values of p and $\partial p/\partial t$ in the interior of $\mathscr{B}$, and has $p=0$, $\partial p/\partial t=0$ elsewhere. Then it follows at once from the uniqueness theorem that $p=0$, $\partial p/\partial t=0$ at all points $(\mathbf{x}, t)$ such that $0 \leqslant t \leqslant \tau_i(\mathbf{x})$. In particular, this holds on $\mathscr{R}$, and so the boundary condition (2.5.1) is satisfied on $\mathscr{R}$ for $t \leqslant \tau_i(\mathbf{x})$. The dependence domain of any point $(\mathbf{x}, t)$ of Σ with $\tau_i(\mathbf{x}) \leqslant t \leqslant \tau_r(\mathbf{x})$ therefore contains only points of $\mathscr{R}$ at which $p=0$, $\partial p/\partial t=0$ and so the disturbance propagated in the extended medium must, by the uniqueness theorem, coincide with the solution of the reflexion problem for $\tau_i(\mathbf{x}) \leqslant t \leqslant \tau_r(\mathbf{x})$. Since this solution is unique, the disturbance defined in this way is also independent of the manner in which the extended medium is defined, and we can speak of it as the incident disturbance without having to specify an unlimited medium in which this 'would propagate in the absence of any obstacles'. What we have shown is that the *secondary disturbance due to the presence of the reflector has as its front the reflected wave front determined by geometrical optics*. This is of course only true in the domain of direct reflexion whose points can be reached along reflected rays. In the shadow, the front of the secondary disturbance coincides with the incident front and the secondary disturbance simply cancels the incident pulse, until the diffracted front arrives.

We can now complete §1.5 in which reflexion boundary conditions were discussed. Let a pulse p_i be incident on a reflector (or a system of reflectors) $\mathscr{R}$, and let p be the total resulting pressure field. One condition to be satisfied by p is of course the boundary condition on $\mathscr{R}$. To obtain a second condition, we can argue as follows. The front of p_i is represented by a characteristic C_i and $\mathscr{R}$ determines a 3-cylinder R parallel to the t-axis. C_i and R meet in a 2-space $\mathscr{R}^*$. We now construct the union of the influence domains of all the points of $\mathscr{R}^*$; it is a domain Σ^* of space-time whose projection is the physical space S. Then the second boundary condition to be satisfied by p is that $p-p_i$ must be zero outside Σ^*. The boundary of Σ^* consists of reflexion characteristics and of portions of C_i. If p is a continuous solution of the wave equation (as has been assumed so far) then $p - p_i$ must vanish on this boundary. In a shadow, the boundary of Σ^* can be replaced by the appropriate diffraction characteristics.

This formulation of the second boundary condition assumes that the secondary front has been determined. There is an alternative formulation which is particularly useful when Laplace transforms are employed. The domain Σ^* has the following property: if $(\mathbf{x}, t)$ is a point of space-time such that $\mathbf{x}$ belongs to the open domain S (physical space), then the dependence domain of $(\mathbf{x}, t)$ meets Σ^* in a bounded domain of space-time. We can therefore say that (i) p satisfies the boundary condition on $\mathscr{R}$ and (ii) $p - p_i$ vanishes outside some domain Σ^* which meets the dependence domain of any point $(\mathbf{x}, t)$, where $\mathbf{x}$ is a point of S, in a bounded domain.† It can be shown that these conditions determine p uniquely and that Σ^* is in fact the domain defined in the preceding paragraph.

APPENDIX

The characteristics containing a given 2-space; caustics

It was shown in § 3 that the characteristics containing a given 2-space $\mathscr{S}$ can be obtained as follows. Let the equation of $\mathscr{S}$ in parametric form be

$$\mathbf{x} = \mathbf{x}_0(\lambda, \mu), \quad t = t_0(\lambda, \mu). \tag{2.A.1}$$

The characteristics containing $\mathscr{S}$ are then made up of the bicharacteristics which are solutions of the differential equations (2.3.7) that reduce to $x_0(\lambda, \mu)$, $t_0(\lambda, \mu)$ for $\nu = 0$; the value of the vector $\mathbf{y} = \mathrm{d}\mathbf{x}/\mathrm{d}\nu$ for $\nu = 0$, that is to say on $\mathscr{S}$, is then determined by the equations (2.3.9) and (2.3.11),

$$\mathrm{d}t_0 = \mathbf{y}_0 \cdot \mathrm{d}\mathbf{x}_0, \tag{2.A.2}$$

$$c_0^2 y_0^2 = 1, \quad c_0 = c\{\mathbf{x}(\lambda, \mu)\}. \tag{2.A.3}$$

Assuming that the equations of the bicharacteristics can be integrated, the problem therefore reduces to the solution of these equations. This can be effected by a simple geometrical construction.

† This can be expressed concisely by saying that $p - p_i$ is compact towards the past (p. 135).

The equations (2.A.1) define a moving curve σ_t in space which describes a fixed surface $\mathscr{S}_0$ whose equation is $\mathbf{x}=\mathbf{x}_0(\lambda, \mu)$. The characteristics that contain $\mathscr{S}$ represent wave fronts which meet $\mathscr{S}_0$ at time t_0 along σ_t. By (2.A.3) and the definition of $\mathbf{y}$, $c_0\mathbf{y}_0$ is the unit vector in the direction of the ray at every point of σ_t. To interpret the condition (2.A.2) we write it as

$$c_0\mathbf{y}_0 . \mathbf{V}_0 = c_0, \tag{2.A.4}$$

where

$$\mathbf{V}_0 = \frac{d\mathbf{x}_0}{dt_0} = \frac{\dfrac{\partial \mathbf{x}_0}{\partial \lambda}d\lambda + \dfrac{\partial \mathbf{x}_0}{\partial \mu}d\mu}{\dfrac{\partial t_0}{\partial \lambda}d\lambda + \dfrac{\partial t_0}{\partial \mu}d\mu}. \tag{2.A.5}$$

Clearly, $\mathbf{V}_0$ is the velocity of a point moving with σ_t which coincides with $\mathbf{x}_0$ at time t_0 and with $\mathbf{x}_0 + d\mathbf{x}_0$ at time $t_0 + dt_0$. $\mathbf{V}_0$ is tangential to $\mathscr{S}_0$ and depends on $d\lambda/d\mu$. The condition (2.A.4) means that $c_0\mathbf{y}_0$ is such that the component of $\mathbf{V}_0$ in the direction of $c_0\mathbf{y}_0$ is c_0, whatever the value of $d\lambda/d\mu$. It follows that $\mathbf{y}_0$, and therefore the characteristics containing $\mathscr{S}$, can be real only if $V_0 \geqslant c_0$ for all values of $d\lambda/d\mu$. Now V_0 is clearly least when $d\lambda/d\mu$ is such that $\mathbf{V}_0$ is normal to σ_t. Let us denote the minimum value of V_0 by v_0.† Then real characteristics containing $\mathscr{S}$ can be constructed if and only if $v_0 \geqslant c_0$ everywhere on $\mathscr{S}$.

The component of $\mathbf{V}_0$ normal to σ_t is always v_0. Hence we can represent $\mathbf{V}_0$ as follows. At a point $\mathbf{x}_0$ of σ_t draw a vector $\mathbf{v}_0$ which is tangent to $\mathscr{S}_0$, normal to σ_t, and whose magnitude is v_0. Through the end-point of this vector draw a line l normal to it and in the tangent plane to $\mathscr{S}_0$ at $\mathbf{x}_0$ (fig. 2.7). Then if $\mathbf{X}$ is a point of l, the vector $\mathbf{V}_0$ in the direction $\mathbf{X}-\mathbf{x}_0$ is simply $\mathbf{X}-\mathbf{x}_0$. Now let $\mathscr{K}$ be a sphere with centre $\mathbf{x}_0$, of radius c_0. The tangent planes to $\mathscr{K}$ through l touch it in general at two points $\mathbf{Y}_1, \mathbf{Y}_2$. Obviously the component of $\mathbf{X}-\mathbf{x}_0$ in either of the directions $\mathbf{Y}_1-\mathbf{x}_0$, $\mathbf{Y}_2-\mathbf{x}_0$ is c_0; hence $\mathbf{Y}_1-\mathbf{x}_0$, $\mathbf{Y}_2-\mathbf{x}_0$ define the two possible directions of the ray. These vectors in fact represent $c_0^2\mathbf{y}_0$. They are both in the plane determined by $\mathbf{v}_0$ and the normal $\mathbf{n}_0$ to $\mathscr{S}_0$ at $\mathbf{x}_0$, and equally inclined at an angle $\cos^{-1}(c_0/v_0)$ to $\mathbf{v}_0$.

If $t_0=0$, $\mathscr{S}$ represents the fixed surface $\mathscr{S}_0$ at time $t=0$ so that the characteristics containing $\mathscr{S}$ represent the wave fronts which reduce to $\mathscr{S}_0$ when $t=0$. In this case v_0 is infinite so that the rays are normal to $\mathscr{S}_0$.

† It is easily shown from (2.A.5) that v_0 is given by

$$\frac{EG-F^2}{v_0^2} = E\left(\frac{\partial t_0}{\partial \mu}\right)^2 - 2F\frac{\partial t_0}{\partial \lambda}\frac{\partial t_0}{\partial \mu} + G\left(\frac{\partial t_0}{\partial \lambda}\right)^2,$$

where E, F, and G are the coefficients of the first differential form of $\mathscr{S}_0$, that is to say

$$|d\mathbf{x}_0|^2 = E\,d\lambda^2 + 2F\,d\lambda\,d\mu + G\,d\mu^2.$$

There is only one family of wave fronts (which consists of the geodesic parallels of $\mathscr{S}_0$), but there are two characteristics representing the two possible opposite directions of propagation.

In the special case where $v_0 = c_0$ at all points of $\mathscr{S}$ there is only one characteristic. The rays are then tangential to $\mathscr{S}_0$. If, in the notation of § 3, $\mathbf{x}(\lambda, \mu, \nu)$ represents the ray system, then the Jacobian

$$J = \mathrm{D}(x_1, x_2, x_3)/\mathrm{D}(\lambda, \mu, \nu)$$

vanishes on $\mathscr{S}_0$. For

$$J = \left(\frac{\partial \mathbf{x}}{\partial \lambda} \wedge \frac{\partial \mathbf{x}}{\partial \mu}\right) \cdot \frac{\partial \mathbf{x}}{\partial \nu}$$

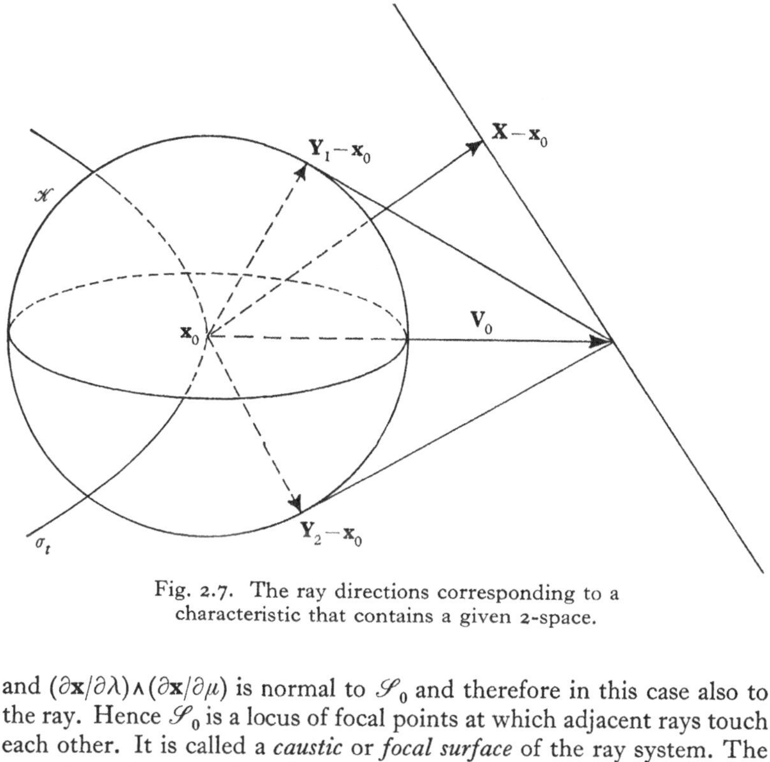

Fig. 2.7. The ray directions corresponding to a characteristic that contains a given 2-space.

and $(\partial\mathbf{x}/\partial\lambda) \wedge (\partial\mathbf{x}/\partial\mu)$ is normal to $\mathscr{S}_0$ and therefore in this case also to the ray. Hence $\mathscr{S}_0$ is a locus of focal points at which adjacent rays touch each other. It is called a *caustic* or *focal surface* of the ray system. The wave fronts have an edge of regression along σ_t.

Conversely, if a ray system $\mathbf{x}(\lambda, \mu, \nu)$ is given then its caustics are determined by the condition $J = 0$. Suppose, for instance, that one of the associated fronts, $\mathscr{W}_0$, is given and corresponds to $\nu = 0$. We must assume that $J \neq 0$ on $\mathscr{W}_0$ and hence by continuity $J \neq 0$ for sufficiently small ν. On each ray, there will be focal points determined by the least values of ν and $-\nu$ for which $J = 0$; their loci are the caustics nearest to $\mathscr{W}_0$. They bound a domain containing $\mathscr{W}_0$ in which the rays form a field (in the language of the calculus of variations); there is one and only one ray through every point of this domain. But a point outside this domain may be on several rays, or on none. This leads to a number of analytical difficulties in the theory of the wave equation. In the case of a homogeneous medium, there are in general just two caustics, and they can be found explicitly. This case will be discussed in more detail in the next chapter.

CHAPTER 3

GEOMETRICAL ACOUSTICS

1. Introduction

The important fact that the characteristics are the natural boundaries between different régimes in space-time has been deduced from the uniqueness theorem. It is therefore established as yet only for solutions of the wave equation with continuous derivatives and remains to be proved for acoustic shock fronts where the pressure is discontinuous. This extension is easily effected and the relevance of geometrical optics to the theory of the wave equation is further underlined by the result that along a ray associated with an acoustic shock front the square of the pressure jump divided by the acoustic impedance ρc varies in accordance with the intensity law of geometrical optics. Acoustic shocks can therefore be treated as a separate subject which has been called *geometrical acoustics* by J. B. Keller. Geometrical acoustics can also be considered as a first approximation which can be improved by means of certain series solutions of the wave equation. The exposition of this theory forms the core of this chapter.

But before the theory of acoustic shocks can be discussed, one must first ask what is to be understood by a discontinuous solution of the wave equation or of the equations of motion. These equations are derived from the physical conservation laws of mass, momentum and energy which are macroscopic laws expressed as integral identities. The transition to differential equations is analytically valid only when the first-order derivatives of the pressure and density variations and of the velocity components are continuous. The simplest way of dealing with discontinuity fronts is to apply the conservation laws directly in their integral form; when other difficulties of a similar kind such as algebraic infinities occur the same procedure may be used. There is, however, much to be said for a unified treatment in which the differential operators are reinterpreted in such a way that their domain includes discontinuous and other singular functions. Such extensions have in any case been introduced in the theory of partial differential equations for different reasons. The approach adopted here will be the interpretation of the wave equation and of the equations of motion in terms of the theory of distributions due to Laurent Schwartz. This has considerable technical advantages as well.

It would not be practicable to attempt a systematic exposition of the theory of distributions in this monograph, and the reader unfamiliar with the subject must be referred to the literature.† We shall, however, explain briefly what a distribution is, and state the principal properties as and when they are required.

A convenient starting point is the following concept of the generalized derivative of a function. Let $f(\mathbf{x}, t)$ be a function that is integrable over any bounded domain, and $\phi(\mathbf{x}, t)$ be any continuous function which vanishes outside a bounded domain. Then the integral $\int f\phi \, d\Sigma$ taken over the whole of space-time has a meaning and is a linear functional, associated with f, over the space of functions ϕ; we denote it by $f.\phi$. Now suppose that f has continuous derivatives, and restrict the admissible functions ϕ so that they also have continuous derivatives. Then for instance

$$\frac{\partial f}{\partial t}.\phi = \int \frac{\partial f}{\partial t}\phi \, d\Sigma = -\int f \frac{\partial \phi}{\partial t} d\Sigma = -f.\frac{\partial \phi}{\partial t}$$

by the four-dimensional divergence theorem, since the integral over the boundary of the domain outside which $\phi = 0$ is zero, ϕ being by hypothesis continuous.‡ Now the formulae

$$\frac{\partial f}{\partial x_\alpha}.\phi = -f.\frac{\partial \phi}{\partial x_\alpha} \quad (\alpha = 1, 2, 3), \quad \frac{\partial f}{\partial t}.\phi = -f.\frac{\partial \phi}{\partial t} \qquad (3.1.1)$$

have a meaning even when the derivatives of f do not exist. These formulae define the *generalized derivatives* of f. They need not be functions in the ordinary sense at all. If for instance $f = H(t)$, that is to say, $f = 1$ $(t > 0)$, $f = 0$ $(t < 0)$, then

$$\frac{\partial H(t)}{\partial t}.\phi = -\int d\mathbf{x} \int_0^\infty \frac{\partial \phi}{\partial t} dt = \int \phi(\mathbf{x}, 0) \, d\mathbf{x}.$$

† The standard work is Schwartz (1950–51). At the time of writing there appears to be no other complete exposition of the theory, except for L. Schwartz's mimeographed lecture notes, *Méthodes mathématiques de la physique* (Centre de documentation universitaire, Paris, 1957). Some aspects of distributions are treated in Friedman (1956). A different approach to the subject, in terms of 'generalized functions', has been developed by Temple; see Temple (1956).

‡ This domain is called the *support* of ϕ. It is properly defined as the closure of the set of points at which $\phi \neq 0$.

This can be written symbolically as

$$H'(t) = \delta(t), \tag{3.1.2}$$

where $\delta(t)$ is the Dirac delta function. Again, if $f = t^{-\frac{1}{2}}H(t)$ then

$$\frac{\partial}{\partial t}\{H(t)\,t^{-\frac{1}{2}}\}\,.\,\phi = -\int \mathrm{d}\mathbf{x}\int_0^\infty \frac{\partial\phi}{\partial t}\,t^{-\frac{1}{2}}\,\mathrm{d}t$$

$$= -\tfrac{1}{2}Pf\int \phi t^{-\frac{3}{2}}H(t)\,\mathrm{d}\Sigma, \tag{3.1.3}$$

where the prefix Pf indicates that the finite part of the divergent integral, in the sense of Hadamard,† is to be calculated.

This process of symbolic differentiation can be continued provided that the admissible functions ϕ are restricted suitably. We shall say that ϕ is a *test function* if it (i) vanishes outside a bounded domain and (ii) has continuous derivatives of all orders. Then provided that all functionals are understood to be over the space of test functions, a function has generalized derivatives of all orders. Any sum of such generalized derivatives is a *distribution*, and all distributions can in effect be obtained in this way.‡ The product of a function f and of a distribution T is defined by $fT.\phi = T.f\phi$ and the rule for the differentiation of a product applies.§ Any linear partial differential equation can therefore be taken to apply to distributions as well as to functions.

The theory of distributions has certain features which make for technical simplicity, but its introduction in a physical subject appears at first sight to be an artificiality. One may, however,

† Hadamard (1923). Accounts of Hadamard's theory are given in several books, for instance, Courant-Hilbert (1937) or Ward (1955).

‡ See Schwartz (1950–51), ch. III. The proper definition of a distribution is that it is a linear functional on the space of test functions. That is to say, a distribution T defines a functional $T.\phi$ which is distributive,

$$T.k_1\phi_1 + T.k_2\phi_2 = T.(k_1\phi_1 + k_2\phi_2),$$

where k_1, k_2 are complex constants, and which is continuous. To define continuity, one must first introduce convergence in the space of test functions; we say that a sequence ϕ_n tends to zero in the space of test functions if the supports of all the ϕ_n are contained in a fixed domain, and the ϕ_n and their derivatives of all orders tend to zero uniformly in $\mathbf{x}$ and t. Then, for a distribution, $\phi_n \to 0$ in this sense implies that also $T.\phi_n \to 0$.

§ Strictly speaking this definition is valid only for indefinitely differentiable functions f. But if T is of order $\leqslant n$, that is to say, $T.\phi$ is defined for all test functions that are n times continuously differentiable, then fT is defined if f is at least n times continuously differentiable.

perhaps justify it as follows. It is usual to think of the physical variables as functions of position in space-time. But this is already a mathematical model; one never measures the pressure, for example, at a particular point and a given instant, but only an average over a small region and a short time-interval. The result of such a measurement can equally well be expressed as $p\,.\,\phi$, where ϕ is a test-function which represents the averaging of the pressure effected by the actual measurement. Thus a mathematical model in which the physical variables are represented by distributions and the observations by the linear functionals which these define is not necessarily more artificial than one in which the physical variables are represented by continuous and differentiable functions defined on a mathematical continuum. The overriding advantage of the latter and more usual scheme is its greater universality, since distributions can only be introduced when the equations are linear. But once a linear theory is assumed to be valid, as in this monograph, there seems to be no inherent reason for clinging to the concepts of classical analysis when an extension such as that furnished by the theory of distributions gives a more flexible theory which is equally easy to handle.

2. Weak solutions of the wave equation

The case with which we shall be concerned most frequently is that of a function that is not everywhere differentiable. If such a function p satisfies the wave equation in the sense of the theory of distributions then by (3.1.1)

$$L(p)\,.\,\phi = p\,.\,L(\phi) = \int pL(\phi)\,\mathrm{d}\Sigma = 0, \tag{3.2.1}$$

where L denotes the differential operator

$$L \equiv \frac{1}{\rho c^2}\frac{\partial^2}{\partial t^2} - \frac{\partial}{\partial x_\alpha}\frac{1}{\rho}\frac{\partial}{\partial x_\alpha}, \tag{3.2.2}$$

and the integral is taken over the whole of space-time. A function which satisfies (3.2.1) is called a *weak solution* of the wave equation. The concept of a weak solution of a partial differential equation or system of such equations was first suggested in Courant-Hilbert (1937). There are various definitions of weak solutions, all using

(3.2.1) but with different classes of admissible functions p and test-functions ϕ. It may be noted that it would be sufficient in (3.2.1) to assume that the test-functions are twice continuously differentiable and vanish outside a bounded domain. Since any such function can be approximated uniformly, together with its derivatives of the first and second order, by a sequence of indefinitely differentiable test-functions our definition entails no restriction. The concept of a weak solution can also be introduced in the non-linear case; applied to the equations of motion of an inviscid gas, it leads to a unified treatment of continuous flow and shock waves (Lax, 1954).

A function p which has continuous derivatives up to the second order at least and satisfies the wave equation will be called a *strict solution* of the wave equation. Now we can prove that if a weak solution p of the wave equation has continuous derivatives up to the second order in a domain Ω of space-time then it is a strict solution in Ω. The proof is an immediate consequence of an integral identity of fundamental importance in the theory of the wave equation, the *adjointness relation* or *Green's formula* of the operator L.

Let p and q be twice continuously differentiable functions. Then

$$qL(p)-pL(q)=\frac{\partial}{\partial t}\left\{\frac{1}{\rho c^2}\left(q\frac{\partial p}{\partial t}-p\frac{\partial q}{\partial t}\right)\right\}-\frac{\partial}{\partial x_\alpha}\left\{\frac{1}{\rho}\left(q\frac{\partial p}{\partial x_\alpha}-p\frac{\partial q}{\partial x_\alpha}\right)\right\}.$$

If we integrate this over a bounded domain Σ of space-time with piece-wise smooth frontier S and transform the right-hand side into an integral over S by means of the four-dimensional divergence theorem, we obtain the Green's formula. We shall write it as

$$\int_\Sigma \{qL(p)-pL(q)\}\,\mathrm{d}\Sigma=\int_S\left(q\frac{\partial p}{\partial l}-p\frac{\partial q}{\partial l}\right)\frac{\mathrm{d}S}{\rho}, \qquad (3.2.3)$$

where $\partial/\partial l$ is called the *transversal derivative* on S and is defined by

$$\frac{\partial}{\partial l}=\frac{n_0}{c^2}\frac{\partial}{\partial t}-n_\alpha\frac{\partial}{\partial x_\alpha}, \qquad (3.2.4)$$

(n_1, n_2, n_3, n_0) being the direction cosines of the normal to S drawn away from Σ. The 4-vector $(-n_1, -n_2, -n_3, n_0/c^2)$ defines the transversal to S. This has the important property that it is tangential to S if, and only if, S is a characteristic. In fact the necessary and sufficient condition for this is

$$\frac{n_0}{c^2} \cdot n_0 - n_\alpha \cdot n_\alpha = 0,$$

which is the equation defining the characteristics. It may be noted that when S is a characteristic then the transversal is at every point in the direction of the bicharacteristic.

Let us now consider a weak solution p of the wave equation which is twice continuously differentiable in a domain Ω. We can then apply (3.2.3) with $q = \phi$, ϕ being any test-function whose support Σ is contained in Ω. The integral over S is zero, since ϕ and its derivatives are continuous, and so it follows from the definition of a weak solution that

$$\int_\Sigma L(p)\, \phi \, d\Sigma = 0.$$

But $L(p)$ is continuous, by hypothesis; hence $L(p) = 0$ in Σ, by the fundamental lemma of the calculus of variations. As Σ may be any sub-domain of Ω, our assertion follows. This lemma will prove very useful. A case important in physical applications is that of a weak solution which is well-behaved, say twice continuously differentiable, everywhere except on certain isolated 3-spaces that carry singularities. Our lemma shows that except on the singularity-carrying 3-spaces the solution will satisfy the wave equation in the strict sense. Such solutions will be referred to as *piece-wise strict solutions*.

3. The propagation of discontinuities

We shall now show that an acoustic shock front is a wave front in the sense of geometrical optics, and that the pressure rise satisfies an ordinary differential equation, called the *transport equation*, on each ray. This is the most important result of the present chapter, and is an essential complement to the theory developed in chapter 2.

It will be sufficient to consider a piece-wise strict solution p of the wave equation which has a simple discontinuity across a 3-space S. We may suppose that S divides space-time into two domains Ω_1 and Ω_2; we assume that p and its first-order derivatives tend to limits which are continuous functions on S as S is approached either from Ω_1 or from Ω_2, but that these limits are different. Let ϕ be any test-function whose support Σ meets S, and let Σ_1 and Σ_2 be the domains formed by the points of Σ which are in Ω_1 and in Ω_2 respectively.† Then

$$0=\int_{\Sigma} pL(\phi)\,d\Sigma=\int_{\Sigma_1} pL(\phi)\,d\Sigma+\int_{\Sigma_2} pL(\phi)\,d\Sigma.$$

Now the Green's formula (3.2.3) can be applied to the integrals over Σ_1 and Σ_2 separately. The contributions from the frontier of the support of ϕ vanish, since ϕ and its derivatives are continuous. By the lemma of § 2 p is a strict solution of the wave equation in Σ_1 and Σ_2. Hence we shall be left with two integrals over S, which we can write jointly as

$$\int_S \left\{\Delta p\left[\frac{\partial\phi}{\partial l}\right]-[\phi]\,\Delta\left(\frac{\partial p}{\partial l}\right)\right\}\frac{dS}{\rho}=0. \tag{3.3.1}$$

Here quantities in square brackets are to be evaluated on S, and the prefix Δ indicates that the saltus across S of the quantity following it is to be taken. The transversal derivative may be formed with the direction cosines of the normal to S pointing into either domain, say into Ω_1; it is because the transversal derivative formed by means of the other normal is $-(\partial/\partial l)$ that the discontinuous variations Δp, $\Delta(\partial p/\partial l)$ appear in (3.3.1).

Suppose first that S is not a characteristic. Then the transversal is not tangential to S. Hence $[\phi]$ and $[\partial\phi/\partial l]$ are independent; if two indefinitely differentiable compact functions ϕ_0 and ϕ_1 are given on S, one can always construct a test-function ϕ such that $[\phi]=\phi_0$ and $[\partial\phi/\partial l]=\phi_1$. Since by hypothesis Δp and $\Delta[\partial p/\partial l]$ are continuous, the fundamental lemma of the calculus of variations implies in this case that $\Delta p=0$, $\Delta(\partial p/\partial l)=0$ on S. Using once more the fact that the transversal is not tangential to S we deduce that p and its first-order derivatives are continuous across S. This contradicts our hypothesis. *Hence a 3-space which carries a simple*

† Strictly speaking, since Σ is by definition closed, Σ_1 and Σ_2 consist of the interior points of Σ in Ω_1 and Ω_2 respectively. It may be noted in passing that the argument to be developed applies in any domain Ω that meets S and is such that p is twice continuously differentiable in $\Omega-S$. The assumption that p is a strict solution everywhere except on S has only been made to simplify the exposition.

discontinuity of p or of its first-order derivatives is necessarily a characteristic.

Suppose now that S is a characteristic. We can take its equation in the form

$$t=\tau(\mathbf{x}), \tag{3.3.2}$$

where τ satisfies the eikonal equation

$$\frac{\partial\tau}{\partial x_\alpha}\frac{\partial\tau}{\partial x_\alpha}=\frac{1}{c^2}. \tag{3.3.3}$$

Now $$\frac{\partial}{\partial x_\alpha}[\phi]=\left[\frac{\partial\phi}{\partial x_\alpha}\right]+\left[\frac{\partial\phi}{\partial t}\right]\frac{\partial\tau}{\partial x_\alpha}\quad(\alpha=1,2,3). \tag{3.3.4}$$

Thus $$\mathrm{d}S\left[\frac{\partial\phi}{\partial l}\right]=n_0\,\mathrm{d}S\left[\frac{1}{c^2}\frac{\partial\phi}{\partial t}+\frac{\partial\tau}{\partial x_\alpha}\frac{\partial\phi}{\partial x_\alpha}\right]$$
$$=\mathrm{d}x\left\{\frac{1}{c^2}\left[\frac{\partial\phi}{\partial t}\right]+\frac{\partial\tau}{\partial x_\alpha}\frac{\partial[\phi]}{\partial x_\alpha}-\frac{\partial\tau}{\partial x_\alpha}\frac{\partial\tau}{\partial x_\alpha}\left[\frac{\partial\phi}{\partial t}\right]\right\},$$

whence by (3.3.3)

$$\mathrm{d}S\left[\frac{\partial\phi}{\partial l}\right]=\mathrm{d}x\frac{\partial\tau}{\partial x_\alpha}\frac{\partial[\phi]}{\partial x_\alpha}. \tag{3.3.5}$$

Since the same calculation applies to the limiting values of p on either side of S we have also

$$\mathrm{d}S\,\Delta\left(\frac{\partial p}{\partial l}\right)=\mathrm{d}x\frac{\partial\tau}{\partial x_\alpha}\frac{\partial}{\partial x_\alpha}\Delta p. \tag{3.3.6}$$

Thus (3.3.1) becomes in this case

$$\int\left\{\Delta p\frac{\partial\tau}{\partial x_\alpha}\frac{\partial}{\partial x_\alpha}[\phi]-[\phi]\frac{\partial\tau}{\partial x_\alpha}\frac{\partial}{\partial x_\alpha}\Delta p\right\}\frac{\mathrm{d}x}{\rho}=0.$$

We can now remove the derivatives of $[\phi]$ by an application of the divergence theorem. The surface integral which arises is over the intersection of the frontier of the support of ϕ and of S, so that it vanishes because of the continuity of ϕ. Hence we find

$$\int\left\{\frac{2}{\rho}\frac{\partial\tau}{\partial x_\alpha}\frac{\partial}{\partial x_\alpha}\Delta p+\Delta p\frac{\partial}{\partial x_\alpha}\left(\frac{1}{\rho}\frac{\partial\tau}{\partial x_\alpha}\right)\right\}[\phi]\,\mathrm{d}x=0,$$

and this implies that Δp satisfies the differential equation

$$\mathrm{Tr}(\tau,\Delta p)=\frac{2}{\rho}\frac{\partial\tau}{\partial x_\alpha}\frac{\partial}{\partial x_\alpha}(\Delta p)+\Delta p\frac{\partial}{\partial x_\alpha}\left(\frac{1}{\rho}\frac{\partial\tau}{\partial x_\alpha}\right)=0 \tag{3.3.7}$$

(provided that the derivatives of τ up to the second order are continuous). This equation is called, following Luneburg, the *transport equation* associated with the wave fronts (3.3.2). Since the $\partial\tau/\partial x_\alpha$ are proportional to the direction cosines of the rays, the transport equation is in effect an ordinary differential equation for Δp along each ray. Its integration, which is elementary, will be considered in § 5.

The same arguments may be used with the equations of motion (1.2.4) and (1.2.9). It is found that

$$\rho n_0 \Delta\mathbf{u} + \mathbf{n}\Delta p = 0, \quad n_0 \Delta p + \rho c^2 \mathbf{n}.\Delta\mathbf{u} = 0, \tag{3.3.8}$$

where $\mathbf{n}$ is vector (n_1, n_2, n_3) and $\Delta\mathbf{u}$ is the discontinuous variation of the velocity across S. In the case of an acoustic shock wave (3.3.2) we have therefore

$$\rho c \Delta\mathbf{u} = c\Delta p \operatorname{grad} \tau. \tag{3.3.9}$$

Now $c \operatorname{grad} \tau$ is the unit vector in the direction of propagation. Hence the velocity discontinuity is longitudinal and its magnitude is $\Delta p/\rho c$. If a velocity potential Φ satisfying (1.3.5) exists then (3.3.9) shows that it is continuous across the shock front. The equations (3.3.8) also admit stationary discontinuities. Then $n_0 = 0$ so that

$$\Delta p = 0, \quad \mathbf{n}.\Delta\mathbf{u} = 0. \tag{3.3.10}$$

Hence the pressure and the normal velocity are continuous and the tangential velocity and the entropy are discontinuous. This is the acoustic analogue of a *contact discontinuity* (a vortex sheet).

To conclude this section we will extend our results to discontinuities of higher order. We shall say that a 3-space S carries a discontinuity of order $m \geqslant 1$ if the derivatives of p of all orders up to $m-1$ are continuous, and those of order m discontinuous, across S. In the neighbourhood of S, p is to be at least $m+1$ times continuously differentiable. Then the strict solution p also satisfies the wave equation in the sense of the theory of distributions. The distribution $\partial^{m-1}p/\partial t^{m-1}$ therefore also satisfies the wave equation. As it is a function, it is a weak solution; its first-order derivatives are discontinuous across S and so *S must be a characteristic.* The saltus of $\partial^m p/\partial t^m$ across S satisfies the transport equation.

4. The propagation of algebraic infinities

Solutions with algebraic infinities sometimes occur in applications. An important example is the elementary solution (see appendix to chapter 5) of the ordinary two-dimensional wave equation,

$$\frac{c}{2\pi}\frac{H(ct-r)}{(c^2t^2-r^2)^{\frac{1}{2}}}, \quad r^2=x_1^2+x_2^2.$$

This is a weak solution in any domain that does not meet the x_3-axis. Such singularities are also always carried by characteristics. (They probably also correspond to shock waves.)

To prove this, consider a piece-wise strict solution p with an algebraic infinity on a 3-space whose equation is

$$S(\mathbf{x},t)=0. \tag{3.4.1}$$

We will suppose that S is twice continuously differentiable and that

$$\frac{\partial S}{\partial x_\alpha}\frac{\partial S}{\partial x_\alpha}+\left(\frac{\partial S}{\partial t}\right)^2 \neq 0$$

on $S=0$, so that there are no singular points. The function S need only be defined in a neighbourhood of the 3-space $S=0$, say in $|S|<M$, and we can take M sufficiently small to ensure that p is a strict solution in $0<|S|<M$. We assume that

$$p=US^{-a}+W \quad (0<S<M), \quad p=V|S|^{-b}+W \quad (0>S>-M), \tag{3.4.2}$$

where U, V and W are twice continuously differentiable and the constants a, b satisfy $0<a<1$, $0<b<1$.

Let ϕ be a test-function whose support Σ is contained in $|S|<M$ and meets $S=0$. Then

$$\int_\Sigma pL(\phi)\,\mathrm{d}\Sigma=0.$$

By the definition of an improper integral this means that

$$\lim_{\epsilon\to+0}\int_{\Sigma_\epsilon} pL(\phi)\,\mathrm{d}\Sigma+\lim_{\delta\to+0}\int_{\Sigma_\delta} pL(\phi)\,\mathrm{d}\Sigma=0, \tag{3.4.3}$$

where Σ_ϵ, Σ_δ are the sub-domains of Σ at which $S>\epsilon$, $S<-\delta$ respectively. Since $L(p)=0$ in Σ_ϵ and ϕ vanishes together with its derivatives on the frontier of its support, it follows from the Green's formula (3.2.3) that

$$\int_{\Sigma_\epsilon} pL(\phi)\,\mathrm{d}\Sigma = \int_{S=\epsilon}\left(p\frac{\partial\phi}{\partial l} - \phi\frac{\partial p}{\partial l}\right)\frac{\mathrm{d}S}{\rho},$$

where the transversal derivative is formed with the direction cosines of the normal to $S=\epsilon$ drawn into $S<\epsilon$. By (3.4.2),

$$p\frac{\partial\phi}{\partial l} - \phi\frac{\partial p}{\partial l} = (U\epsilon^{-a}+W)\frac{\partial\phi}{\partial l} - \phi\left(-aU\frac{\partial S}{\partial l}\epsilon^{-a-1} + \epsilon^{-a}\frac{\partial U}{\partial l} + \frac{\partial W}{\partial l}\right)$$

on $S=\epsilon$. Hence

$$\int_{\Sigma_\epsilon} pL(\phi)\,\mathrm{d}\Sigma = \epsilon^{-a-1}\int_{S=0} a\phi U\frac{\partial S}{\partial l}\frac{\mathrm{d}S}{\rho} + O(\epsilon^{-a}),$$

where $\partial/\partial l$ now refers to $S=0$ and involves the normal to $S=0$ drawn into $S<0$. Since the corresponding transversal on $S=-\delta$ becomes $-\partial/\partial l$ as $\delta\to 0$, we have similarly that

$$\int_{\Sigma_\delta} pL(\phi)\,\mathrm{d}\Sigma = -\delta^{-b-1}\int_{S=0} b\phi V\frac{\partial S}{\partial l}\frac{\mathrm{d}S}{\rho} + O(\delta^{-b}).$$

Hence (3.4.3) can hold only if

$$\int_{S=0}\phi U\frac{\partial S}{\partial l}\frac{\mathrm{d}S}{\rho} = 0, \quad \int_{S=0}\phi V\frac{\partial S}{\partial l}\frac{\mathrm{d}S}{\rho} = 0.$$

and since ϕ is arbitrary and the integrands are continuous we must have

$$\left[\frac{\partial S}{\partial l}\right]_{S=0} = 0. \tag{3.4.4}$$

But
$$\left[\frac{\partial S}{\partial l}\right]_{S=0} = \left[\frac{n_0}{c^2}\frac{\partial S}{\partial t} - n_\alpha\frac{\partial S}{\partial x_\alpha}\right]_{S=0},$$

and $\partial S/\partial x_\alpha = \Lambda n_\alpha$, $\partial S/\partial t = \Lambda n_0$, where by hypothesis $\Lambda \neq 0$. Thus (3.4.4) implies

$$\frac{n_0^2}{c^2} = n_\alpha n_\alpha,$$

that is to say, $S=0$ must be a characteristic. The same argument applies if S^{-a} is replaced by $\log S$ or S^{-b} is replaced by $\log(-S)$.

Consider a portion of S which is met once (at most) by any parallel to the t-axis. Then $S=0$ has a unique solution $t=\tau(\mathbf{x})$, where τ satisfies the eikonal equation. Clearly

$$S = S_t\{\mathbf{x}, \tau(\mathbf{x})\}\,T + O(T^2), \tag{3.4.5}$$

where
$$T = t - \tau(\mathbf{x}). \tag{3.4.6}$$
Hence (3.4.2) becomes
$$\left.\begin{aligned} p &= F(\mathbf{x}, T)\, T^{-a} + H \quad (T > 0),\\ p &= G(\mathbf{x}, T)(-T)^{-b} + H \quad (T < 0),\end{aligned}\right\} \tag{3.4.7}$$
where F, G and H are again regular (strictly speaking this is only true if S is, say, four times continuously differentiable). Then both $T=\epsilon$ and $T=-\delta$ are characteristics. We can take Σ_ϵ, Σ_δ to be the domains in which $T>\epsilon$, $T<-\delta$ respectively and transform (3.4.3) by the Green's formula into
$$\lim_{\epsilon\to 0+}\int_{T=\epsilon}\left(p\frac{\partial\phi}{\partial l} - \phi\frac{\partial p}{\partial l}\right)\frac{dS}{\rho} + \lim_{\delta\to 0+}\int_{T=-\delta}\left(p\frac{\partial\phi}{\partial l} - \phi\frac{\partial p}{\partial l}\right)\frac{dS}{\rho} = 0. \tag{3.4.8}$$
Both the integrands can be evaluated by means of (3.3.5) and (3.4.7). It is easily shown that as a result of this calculation, (3.4.8) goes into
$$\begin{aligned}\lim_{\epsilon\to 0+}&\left\{\epsilon^{-a}\int\left([\phi]_{T=0}\frac{\partial\tau}{\partial x_\alpha}\frac{\partial F(\mathbf{x},0)}{\partial x_\alpha} - F(\mathbf{x},0)\frac{\partial\tau}{\partial x_\alpha}\frac{\partial}{\partial x_\alpha}[\phi]_{T=0}\right)\frac{d\mathbf{x}}{\rho}\right.\\ &\left.+O(\epsilon^{1-a})\right\} + \lim_{\delta\to 0+}\left\{\delta^{-b}\int\left(G(\mathbf{x},0)\frac{\partial\tau}{\partial x_\alpha}\frac{\partial}{\partial x_\alpha}[\phi]_{T=0}\right.\right.\\ &\left.\left.-[\phi]_{T=0}\frac{\partial\tau}{\partial x_\alpha}\frac{\partial}{\partial x_\alpha}G(\mathbf{x},0)\right)\frac{d\mathbf{x}}{\rho} + O(\delta^{1-b})\right\} = 0.\end{aligned}$$
This can hold only if both integrals vanish separately. Partial integration then shows, as in the similar case considered in § 3, that $F(\mathbf{x},0)$ and $G(\mathbf{x},0)$ satisfy the transport equation associated with the wave fronts $t=\tau(\mathbf{x})$.

The results which we have just obtained can be extended to all values of a and b in (3.4.2) by differentiation, provided that for $a \geqslant 1$ or $b \geqslant 1$ the integral in the equation defining a weak solution is interpreted as the finite or logarithmic part of a divergent integral.

5. Geometrical acoustics

Let us now consider the transport equation
$$\mathrm{Tr}(\tau, P) = \frac{2}{\rho}\frac{\partial\tau}{\partial x_\alpha}\frac{\partial P}{\partial x_\alpha} + P\frac{\partial}{\partial x_\alpha}\left(\frac{1}{\rho}\frac{\partial\tau}{\partial x_\alpha}\right) = 0. \tag{3.5.1}$$

Here P may be the pressure jump Δp at an acoustic shock front, or one of the coefficients of $(t-\tau)^{-a}$, $(t-\tau)^{-b}$ at a singularity front. To integrate this equation, we multiply it by P so that it becomes

$$\operatorname{div}\left(\frac{P^2}{\rho}\operatorname{grad}\tau\right)=0. \tag{3.5.2}$$

Let $\mathscr{S}_0$ be a domain on a wave front $\tau(\mathbf{x})=t_0$ bound by a curve γ_0. The rays through γ_0 constitute a 'ray-tube' $\mathscr{J}$ which intercepts on any other wave front $\tau(\mathbf{x})=t$ a domain $\mathscr{S}$ that corresponds to $\mathscr{S}_0$. Let us integrate (3.5.2) over the domain of space bounded by $\mathscr{S}_0$, $\mathscr{S}$ and $\mathscr{J}$. Then by the divergence theorem

$$\int_{\mathscr{S}_0}\frac{P^2}{\rho}\frac{\partial\tau}{\partial n}\,d\mathscr{S}+\int_{\mathscr{S}}\frac{P^2}{\rho}\frac{\partial\tau}{\partial n}\,d\mathscr{S}+\int\frac{P^2}{\rho}\frac{\partial\tau}{\partial n}\,d\mathscr{S}=0.$$

Now $c\operatorname{grad}\tau$ is the unit vector in the direction of the ray; hence $\partial\tau/\partial n=0$ on $\mathscr{J}$ and $\partial\tau/\partial n=1/c$ on $\mathscr{S}$, $\partial\tau/\partial n=-1/c$ on $\mathscr{S}_0$, if $t>t_0$ (for $t<t_0$ the signs must be reversed). Thus

$$\int_{\mathscr{S}}\frac{P^2}{\rho c}\,d\mathscr{S}=\int_{\mathscr{S}_0}\frac{P^2}{\rho c}\,d\mathscr{S}, \tag{3.5.3}$$

that is to say, the integral of $P^2/\rho c$ over a normal cross-section is constant along a ray-tube. In other words, $P^2/\rho c$ *satisfies the intensity law of geometrical optics* (J. B. Keller, 1954; Friedlander, 1942*a*). Thus we have the important result that *in the acoustic approximation both the location and the strength of a shock can be determined by geometrical optics without reference to the flow on either side of it.* The theory of acoustic shocks can therefore be considered apart from the integration of the wave equation as a separate subject, *geometrical acoustics* (J. B. Keller).

An important example is the case of an acoustic shock wave advancing into an undisturbed medium. At the front there is a sudden pressure rise P, and a velocity U normal to the wave front is suddenly imposed on the medium. By (3.3.9), $U=P/\rho c$, so that the rate at which the shock does work on the medium ahead of it is $PU=P^2/\rho c$. Hence $PU\,d\mathscr{S}$ is constant along any elementary ray-tube; one may say that *the energy of the shock is propagated along ray-tubes.*

The actual law of variation of P along a ray can be obtained from (3.5.3) by making the tube shrink to a ray. We then find

$$P = P_0\left(\frac{\rho c}{\rho_0 c_0}\right)^{\frac{1}{2}}\left(\frac{\mathrm{d}\mathscr{S}_0}{\mathrm{d}\mathscr{S}}\right)^{\frac{1}{2}}, \tag{3.5.4}$$

where ρ_0, c_0, $\mathrm{d}\mathscr{S}_0$ are evaluated at some convenient reference point $\mathbf{x}_0$ on the ray. There is, however, one important limitation; the intensity law (3.5.4) holds only if the arc of the ray joining the reference point to the point $\mathbf{x}$ at which P is calculated *contains no focal points*. This is always the case when $\mathbf{x}$ is sufficiently close to $\mathbf{x}_0$. If $\mathbf{x}$ is moved away from $\mathbf{x}_0$ it may approach a focal point, that is to say, a point at which the ray touches a caustic. Since adjacent rays touch at such a point, we shall have $\mathrm{d}\mathscr{S}/\mathrm{d}\mathscr{S}_0 \to 0$, whence $P/P_0 \to \infty$. Beyond the focal point, (3.5.4) cannot be applied. We shall return to this question in the next section and in the appendix to this chapter.

The formula (3.5.4) can be put into an alternative form. This is best derived directly from the transport equation. Let $\mathbf{x} = \mathbf{x}(\lambda, \mu, \nu)$ be the equations of the rays associated with the wave fronts $t = \tau(\mathbf{x})$, where λ and μ are fixed on each ray, and ν is such that

$$\frac{\partial x_\alpha}{\partial \nu} = \frac{\partial \tau}{\partial x_\alpha} \quad (\alpha = 1, 2, 3). \tag{3.5.5}$$

Then we can make use of a well-known property of the solutions of systems of ordinary differential equation. Let J denote the Jacobian

$$J = \frac{\mathrm{D}(x_1, x_2, x_3)}{\mathrm{D}(\lambda, \mu, \nu)}. \tag{3.5.6}$$

Then

$$\frac{\partial J}{\partial \nu} = \begin{vmatrix} \dfrac{\partial^2 x_1}{\partial\lambda\,\partial\nu} & \dfrac{\partial^2 x_1}{\partial\mu\,\partial\nu} & \dfrac{\partial^2 x_1}{\partial\nu^2} \\ \dfrac{\partial x_2}{\partial\lambda} & \dfrac{\partial x_2}{\partial\mu} & \dfrac{\partial x_2}{\partial\nu} \\ \dfrac{\partial x_3}{\partial\lambda} & \dfrac{\partial x_3}{\partial\mu} & \dfrac{\partial x_3}{\partial\nu} \end{vmatrix} + \dots.$$

But if ξ denotes any of the parameters λ, μ, ν then differentiation of (3.5.5) gives the 'variational equations'

$$\frac{\partial^2 x_\alpha}{\partial\xi\,\partial\nu} = \frac{\partial^2 \tau}{\partial x_\alpha\,\partial x_\gamma}\frac{\partial x_\gamma}{\partial \xi},$$

and if these are substituted in $\partial J/\partial\nu$ it follows at once that

$$\frac{1}{J}\frac{\partial J}{\partial\nu}=\nabla^2\tau. \tag{3.5.7}$$

Hence the transport equation (3.5.1) becomes in terms of λ, μ and ν

$$\frac{2}{\rho}\frac{\partial P}{\partial\nu}+\frac{P}{\rho J}\frac{\partial J}{\partial\nu}-\frac{P}{\rho^2}\frac{\partial\rho}{\partial\nu}=\frac{P}{\rho}\frac{\partial}{\partial\nu}\log\frac{P^2J}{\rho}=0,$$

whence

$$P=P_0\left(\frac{P}{\rho_0}\frac{J_0}{J}\right)^{\frac{1}{2}}, \tag{3.5.8}$$

where the suffix o again refers to a reference point on each ray.

6. Geometrical acoustics in a homogeneous medium

When the medium is homogeneous, the theory of geometrical acoustics can be developed in greater detail. By (2.3.7), the rays are then straight lines; a family of wave fronts consists of parallel surfaces whose common normals are the rays. The function $\tau(x)$ can only be calculated explicitly in a few simple cases, but a parametric representation is always available.

To derive this, let

$$\mathbf{x}=\mathbf{X}(\lambda,\mu) \tag{3.6.1}$$

be the equation of a surface $\mathscr{S}_0$ which is a wave front at some instant, say when $t=0$. One can always choose the order of λ and μ so that the unit vector normal to $\mathscr{S}_0$ given by

$$\mathbf{n}(\lambda,\mu)=\frac{\mathbf{X}_\lambda\wedge\mathbf{X}_\mu}{|\mathbf{X}_\lambda\wedge\mathbf{X}_\mu|} \tag{3.6.2}$$

is directed in the sense of propagation of the fronts (suffixes here denote partial derivatives). It is now convenient to make a slight change of notation by letting ν stand for the distance measured along a ray, so that the position of the front at time t is given by

$$ct=\nu; \tag{3.6.3}$$

it is the surface $\mathscr{S}_\nu$ parallel to, and at a normal distance ν from, $\mathscr{S}_0$. Its equations are

$$\mathbf{x}=\mathbf{X}(\lambda,\mu)+\nu\mathbf{n}(\lambda,\mu). \tag{3.6.4}$$

These equations also represent the ray through $\mathbf{X}(\lambda,\mu)$.

One can always take λ and μ such that the curves $\lambda=\text{const.}$, $\mu=\text{const.}$ are curvature lines on $\mathscr{S}_0$. Then

$$R\mathbf{n}_\lambda=\mathbf{X}_\lambda, \quad S\mathbf{n}_\mu=\mathbf{X}_\mu, \tag{3.6.5}$$

where R and S are the principal radii of curvature of $\mathscr{S}_0$ at $\mathbf{X}$. Hence (3.6.4) gives

$$\mathrm{d}\mathbf{x}=\left(1+\frac{\nu}{R}\right)\mathbf{X}_\lambda \mathrm{d}\lambda+\left(1+\frac{\nu}{S}\right)\mathbf{X}_\mu \mathrm{d}\mu+\mathbf{n}\,\mathrm{d}\nu. \tag{3.6.6}$$

These equations describe a bundle of rays adjacent to the ray (λ, μ). Obviously, an adjacent ray $(\lambda+\mathrm{d}\lambda, \mu)$ meets this ray when $\nu=-R$, while an adjacent ray $(\lambda, \mu+\mathrm{d}\mu)$ meets it when $\nu=-S$; these are the two focal points on the ray, where it touches a caustic. There are therefore in general two caustics. If $R=S$ for all λ, μ then the $\mathscr{S}_\nu$ are concentric spheres, and the caustics degenerate into a single point, the common centre of the spheres. If the S_ν are developable surfaces, one of the radii of curvature is infinite so that there is only a single caustic at finite distance.

It follows from (3.6.6) that $\lambda=\text{const.}$ and $\mu=\text{const.}$ are also curvature lines on $\mathscr{S}_\nu$, and that the principal radii of curvature of $\mathscr{S}_\nu$ are $R+\nu$, $S+\nu$. Now (3.6.5) implies that the curvature lines $\mu=\text{const.}$, $\lambda=\text{const.}$ are convex or concave towards the side of $\mathscr{S}_0$ into which $\mathbf{n}$ points according as $R \gtrless 0$, $S \gtrless 0$ respectively. If we let ν vary from $-\infty$ to ∞ we find therefore that, along a ray, the fronts are in turn concave, anticlastic and convex respectively. The shape of the ray bundle is as shown in fig. 3.1; rays adjacent to (λ, μ) meet two straight lines through the focal points F', F'' orthogonal to $\mathbf{n}$ and to each other.

The element of area on $\mathscr{S}_\nu$ is, by (3.6.6),

$$\mathrm{d}\mathscr{S}_\nu=\left(1+\frac{\nu}{R}\right)\left(1+\frac{\nu}{S}\right)|\mathbf{X}_\lambda \wedge \mathbf{X}_\mu|\,\mathrm{d}\lambda\,\mathrm{d}\mu.$$

As the medium is homogeneous, the density is constant and so the variation of intensity along a ray is given by the *divergence factor*

$$A=\left(\frac{\mathrm{d}\mathscr{S}_0}{\mathrm{d}\mathscr{S}_\nu}\right)^{\frac{1}{2}}=\left\{\frac{RS}{(R+\nu)(S+\nu)}\right\}^{\frac{1}{2}}. \tag{3.6.7}$$

The fronts of a two-dimensional disturbance are parallel

cylinders, and the x_3-axis can be taken parallel to their generators. The equations (3.6.1) then become

$$x_1 = X_1(\lambda), \quad x_2 = X_2(\lambda), \quad x_3 = \mu. \tag{3.6.8}$$

The parameter λ can now be taken to be the arc length on the curves $\mu = \text{const.}$, so that

$$X_1'^2 + X_2'^2 = 1, \quad n_1 = X_2', \quad n_2 = -X_1', \tag{3.6.9}$$

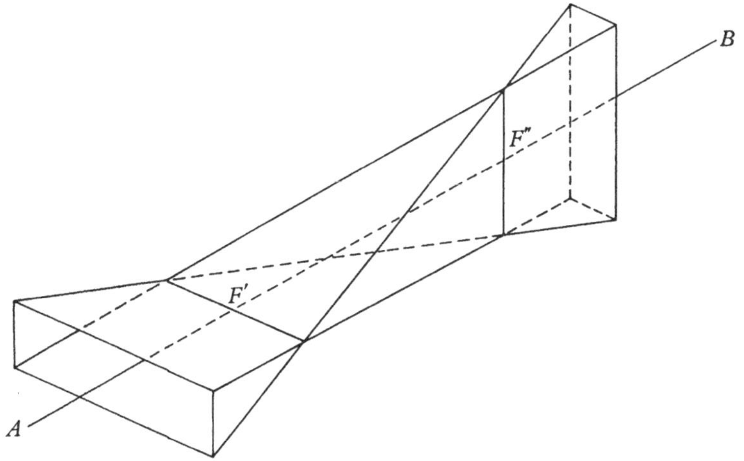

Fig. 3.1. Ray-tube adjacent to a ray AB with two focal points F', F''.

and (3.6.5) then become

$$n_1' = X_2'' = \frac{X_1'}{R}, \quad n_2' = -X_1'' = \frac{X_2'}{R}. \tag{3.6.10}$$

The curves $\mu = \text{const.}$ are given by

$$x_1 = X_1 + \nu X_2', \quad x_2 = X_2 - \nu X_1', \tag{3.6.11}$$

and we have, instead of (3.6.6),

$$dx_1 = \left(1 + \frac{\nu}{R}\right) X_1' d\lambda + X_2' d\nu, \quad dx_2 = \left(1 + \frac{\nu}{R}\right) X_2' d\lambda - X_1' d\nu. \tag{3.6.12}$$

The divergence factor is now

$$A = \left(\frac{R}{R + \nu}\right)^{\frac{1}{2}}. \tag{3.6.13}$$

It is not difficult to determine the shape of the wave front near the caustic in the two-dimensional case. The focal point on the ray

$\lambda = 0$ is given by $\nu = -R(0) = -R_0$ (say), and the equation of the front which touches the caustic at this point is

$$x_1 = X_1(\lambda) - R_0 X_2'(\lambda), \quad x_2 = X_2(\lambda) + R_0 X_1'(\lambda).$$

Let the origin and the direction of the axes be chosen such that

$$X_1'(0) = 0, \quad X_2'(0) = 1; \qquad X_1(0) = R_0, \quad X_2(0) = 0.$$

Then by (3.6.10),

$$x_1(0) = x_2(0) = 0; \quad x_1'(0) = x_2'(0) = 0; \quad x_1''(0) = 0, \quad x_2''(0) = R'(0)/R_0,$$

and so the equation of the front for small λ is, if $R'(0) \neq 0$,

$$x_1 = a\lambda^3 + \ldots, \quad x_2 = b\lambda^2 + \ldots,$$

where a, b are constants. Thus in the neighbourhood of the focal point

$$x_2^3 = Kx_1^2 + \ldots; \tag{3.6.14}$$

the wave front has an edge of regression at the caustic.

A typical example is shown in fig. 3.2, where the initial front $\mathscr{S}_0$ is the parabola $x_2^2 = 2x_1$. The equation of the caustic is then easily shown to be

$$x_2^2 = \{\tfrac{2}{3}(x_1 - 1)\}^3.$$

For $0 \leqslant \nu < 1$ there is no focal point on any ray. For $\nu > 1$, the wave front 'folds over' at the two points A, B where it touches the caustic. If it is assumed that the interior, $x_2^2 < 2x_1$, of the parabola is initially undisturbed, then the undisturbed region is bounded, for $\nu > 1$, by the arcs CD, CE of the front; AB, AC and BC are still parts of the front geometrically, but they are situated in the disturbed region. If $\mathscr{S}_0$ is an acoustic shock front, then the pressure jump is given by the divergence factor formula along ACE and BDE, so that the actual disturbance experienced at a point may involve two shock waves; there is also a singularity on AB, the character of which is investigated in an appendix.

In the three-dimensional case, similar results are obtained, but the analysis is more involved and will be omitted. The wave fronts have an edge of regression where they touch either sheet of the caustic. When there is axial symmetry, one of the sheets of the caustic degenerates into the axis of symmetry, and instead of an edge of regression one obtains a singular point. For instance, if $\mathscr{S}_0$ is the paraboloid of revolution

$$x_2^2 + x_3^2 = 2x_1,$$

then the meridional sections of the fronts are still given by fig. 3.2. But now the x_1-axis is a caustic for $x_1 \geqslant 1$, and the divergence factor formula applies only on CD, CE. On a ray to a point of AC or BC there is one focal point (where the ray meets the x_1-axis), while there are two focal points on a ray to a point of AB, one where it meets the x_1-axis, and one where it touches the caustic

$$x_2^2 + x_3^2 = \{\tfrac{2}{3}(x_1 - 1)\}^3.$$

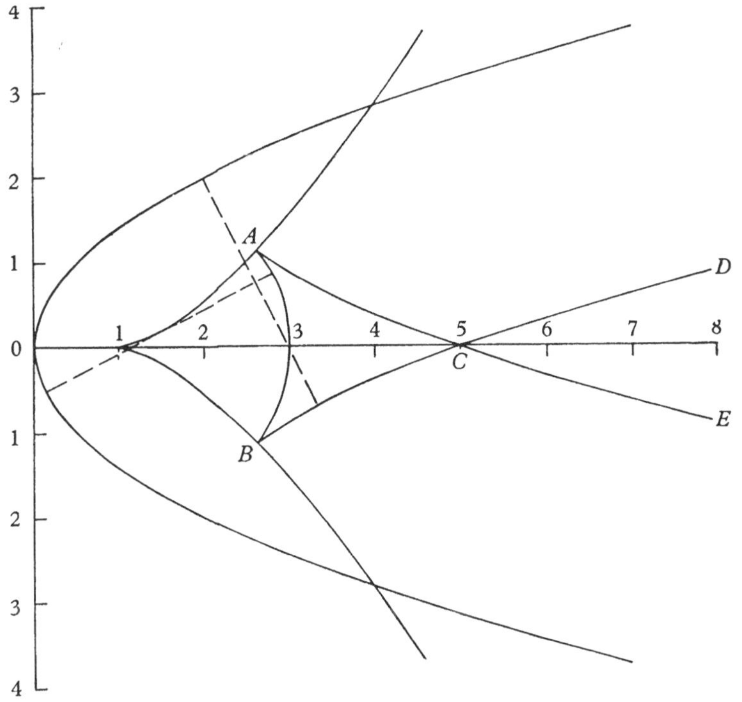

Fig. 3.2. Wave front arising from an initially parabolic front.

It is easily shown that the arrival time v/c is a minimum on CE, CD; a minimax on AC, BC; and a maximum on AB. Similar statements apply in the general case.

What happens to an acoustic shock wave when it passes through a caustic can be investigated by means of Poisson's solution (1.6.2) of the initial-value problem. It is found that a discontinuity is converted into a logarithmic singularity; if there is a second focal point then the logarithmic singularity reverts to a simple discontinuity linked to the initial pressure jump by the geometrical optics intensity law, but with reversed sign. It is not easy to see what the physical significance of these results is. The pressure rise at an acoustic shock front becomes large as a caustic is approached. This implies of course that the linear theory breaks down. It is therefore dubious whether any valid inferences can be drawn from the logarithmic infinity predicted by the linear theory. This itself can at best be taken to be an indication that a shock wave of perhaps moderate strength occurs after focusing, but it gives no quantitative estimate of its magnitude. The result that the simple discontinuity is restored with reversed sign after the second focal point must certainly be rejected. For this discontinuity would be an expansion wave, and it is well known that a discontinuous expansion is always replaced by a continuous transition whose length increases as the pulse propagates. One possible application of the results obtained is to the evaluation of the asymptotic behaviour of a harmonic wave train by Luneburg's method (§ 9); the singularities of the pulse solution give rise to addition terms in the asymptotic expansion of the Fourier-Stieltjes transform (3.9.13). The results obtained can, however, be applied to a discontinuity of higher order.

7. The transport equations of higher order

At an acoustic shock front not only the pressure but also its derivatives are discontinuous. The discontinuous variations of the $\partial^n p/\partial t^n$ $(n = 1, 2, \ldots)$ satisfy certain recurrence relations, the transport equations of higher order. To obtain these we shall first derive the relation between p and $\partial p/\partial t$ which holds on any characteristic. The fact that such a relation exists is a classical property of characteristics. If p is given on a 3-space S, say $p = p_0(\mathbf{x})$ on $t = \sigma(\mathbf{x})$, then the $\partial p/\partial x_\alpha$ on S can be expressed in terms of p_0 and of $p_1(\mathbf{x}) = [\partial p/\partial t]_S$, since

$$\frac{\partial p_0}{\partial x_\alpha} = \left[\frac{\partial p}{\partial x_\alpha}\right]_S + p_1 \frac{\partial \sigma}{\partial x_\alpha} \quad (\alpha = 1, 2, 3).$$

Similarly, all the derivatives of the second order except $[\partial^2 p/\partial t^2]_S$ can be expressed in terms of p_0 and p_1. If p is to be a solution of the wave equation then in general that equation itself serves to determine $[\partial^2 p/\partial t^2]_S$. One can continue this calculation and obtain all the derivatives of p on S in terms of p_0 and p_1, using the equations derived from the wave equation by differentiation. Hence the Taylor series expansion of p in the neighbourhood of any point of S can be found. If ρ, c and the data σ, p_0, p_1 are holomorphic functions of the x_α, then this series converges near the point and solves the Cauchy problem relative to S locally; this is the classical Cauchy-Kowalewsky existence theorem.

But if S is a characteristic then the wave equation does not determine $[\partial^2 p/\partial t^2]_S$. It becomes instead an intrinsic relation on S involving p_0 and p_1 only. The Cauchy problem cannot even be set on S. On the other hand, S may be a 3-space separating different régimes represented by different (analytic) solutions. The correspondence between characteristics and wave fronts can thus be deduced.† The method of Zaremba set out in chapter 2, which is more general (as it does not assume analyticity), supersedes this approach to the theory of wave fronts. But the relation between p and $\partial p/\partial t$ is of considerable importance.

It will be sufficient for our purpose to consider a characteristic whose equation can be put into the form $t=\sigma(\mathbf{x})$, where σ is a solution of the eikonal equation. Then

$$\operatorname{grad}[p]=[\operatorname{grad} p]+\left[\frac{\partial p}{\partial t}\right]\operatorname{grad}\sigma,$$

where quantities enclosed in square brackets are to be evaluated for $t=\sigma(\mathbf{x})$. A further differentiation gives

$$\operatorname{div}\left(\frac{1}{\rho}\operatorname{grad}[p]\right)=\left[\operatorname{div}\left(\frac{1}{\rho}\operatorname{grad} p\right)\right]+\frac{2}{\rho}\operatorname{grad}\sigma.\left[\operatorname{grad}\frac{\partial p}{\partial t}\right]$$
$$+\frac{1}{\rho}\left[\frac{\partial^2 p}{\partial t^2}\right]|\operatorname{grad}\sigma|^2+\left[\frac{\partial p}{\partial t}\right]\operatorname{div}\left(\frac{1}{\rho}\operatorname{grad}\sigma\right).$$

Similarly $$\operatorname{grad}\left[\frac{\partial p}{\partial t}\right]=\left[\operatorname{grad}\frac{\partial p}{\partial t}\right]+\left[\frac{\partial^2 p}{\partial t^2}\right]\operatorname{grad}\sigma.$$

We now eliminate $[\operatorname{grad}\partial p/\partial t]$ between these two equations:

$$\operatorname{div}\left(\frac{1}{\rho}\operatorname{grad}[p]\right)=\left[\operatorname{div}\left(\frac{1}{\rho}\operatorname{grad} p\right)\right]+\frac{2}{\rho}\operatorname{grad}\sigma.\operatorname{grad}\left[\frac{\partial p}{\partial t}\right]$$
$$+\left[\frac{\partial p}{\partial t}\right]\operatorname{div}\left(\frac{1}{\rho}\operatorname{grad}\sigma\right)-\frac{1}{\rho}\left[\frac{\partial^2 p}{\partial t^2}\right]|\operatorname{grad}\sigma|^2.$$

† The classical exposition of this theory will be found in Hadamard (1903).

Since σ satisfies the eikonal equation and p the wave equation, it follows that

$$\frac{2}{\rho}\operatorname{grad}\sigma\,.\operatorname{grad}\left[\frac{\partial p}{\partial t}\right]+\left[\frac{\partial p}{\partial t}\right]\operatorname{div}\left(\frac{1}{\rho}\operatorname{grad}\tau\right)=\operatorname{div}\left(\frac{1}{\rho}\operatorname{grad}[p]\right). \tag{3.7.1}$$

This is the intrinsic form of the wave equation on a characteristic.

Now let $t=\tau(\mathbf{x})$ be the equation of an acoustic shock front. Then $\tau(\mathbf{x})+\epsilon$, $\tau(\mathbf{x})-\delta$ are also solutions of the eikonal equation. Hence we can put $\sigma=\tau+\epsilon$, $\sigma=\tau-\delta$ in (3.7.1), subtract the resulting equations and make $\epsilon\to+0$, $\delta\to+0$, to obtain

$$\frac{2}{\rho}\operatorname{grad}\tau\,.\operatorname{grad}\left(\Delta\frac{\partial p}{\partial t}\right)+\Delta\left(\frac{\partial p}{\partial t}\right)\operatorname{div}\left(\frac{1}{\rho}\operatorname{grad}\tau\right)=\operatorname{div}\left(\frac{1}{\rho}\operatorname{grad}\Delta p\right). \tag{3.7.2}$$

Let us now assume for simplicity that ρ and c are indefinitely differentiable and that p is indefinitely differentiable also on either side of the shock front. Then $L(p)=0$ implies that also $L(\partial^n p/\partial t^n)=0$ for $n=1, 2, \ldots$. Hence (3.7.2) holds if Δp and $\Delta(\partial p/\partial t)$ are replaced by $\Delta(\partial^n p/\partial t^n)$ and $\Delta(\partial^{n+1}p/\partial t^{n+1})$ respectively. If we put

$$\Delta\left(\frac{\partial^n p}{\partial t^n}\right)=\left[\frac{\partial^n p}{\partial t^n}\right]_{\tau-0}^{\tau+0}=P_n(\mathbf{x})\quad(n=0, 1, 2, \ldots), \tag{3.7.3}$$

the resulting *transport equations of higher order* become

$$\frac{2}{\rho}\operatorname{grad}\tau\,.\operatorname{grad}P_n+P_n\operatorname{div}\left(\frac{1}{\rho}\operatorname{grad}\tau\right)=\operatorname{div}\left(\frac{1}{\rho}\operatorname{grad}P_{n-1}\right)\quad(n=1, 2, \ldots), \tag{3.7.4}$$

or, more briefly,

$$\operatorname{Tr}(\tau, P_n)=-L(P_{n-1})\quad(n=1, 2, \ldots). \tag{3.7.5}$$

The transport equations of higher order can be integrated by the procedure adopted for the integration of the transport equation of order zero in § 5. In terms of the curvilinear coordinates λ, μ, ν introduced there, and of the Jacobian J defined by (3.5.6), the equations (3.7.5) are

$$2\frac{\partial P_n}{\partial \nu}-\frac{P_n}{\rho}\frac{\partial\rho}{\partial\nu}+\frac{P_n}{J}\frac{\partial J}{\partial \nu}=\begin{cases}0 & (n=0),\\ -\rho L[P_{n-1}] & (n=1, 2, \ldots),\end{cases}$$

or

$$\frac{\partial}{\partial\nu}\left\{P_n\left(\frac{J}{\rho}\right)^{\frac{1}{2}}\right\}=\begin{cases}0 & (n=0),\\ -\tfrac{1}{2}(\rho J)^{\frac{1}{2}}L[P_{n-1}] & (n=1, 2,\ldots).\end{cases} \tag{3.7.6}$$

Thus the transport equations are ordinary differential equations along each ray associated with the fronts $\tau(\mathbf{x})=t$. Each P_n involves an arbitrary function of λ and μ; to fix this, let us suppose that $\nu=0$ is a surface $\mathscr{S}$ in space on which the P_n are given, say,

$$[P_n]_{\nu=0}=\pi_n(\lambda,\mu), \tag{3.7.7}$$

and that every ray meets $\mathscr{S}$ just once. Then (3.7.6) gives, on integration,

$$P_0=\left(\frac{\rho J_0}{J\rho_0}\right)^{\frac{1}{2}}\pi_0(\lambda,\mu), \tag{3.7.8}$$

$$P_n=\left(\frac{\rho J_0}{J\rho_0}\right)^{\frac{1}{2}}\pi_n(\lambda,\mu)-\frac{1}{2}\left(\frac{\rho}{J}\right)^{\frac{1}{2}}\int_0^\nu(\rho' J')^{\frac{1}{2}}L[P'_{n-1}]\,d\nu', \tag{3.7.9}$$

where ρ_0, J_0 are the values of ρ and of J for $\nu=0$ respectively, and ν is replaced by ν' in ρ', J' and P'_{n-1}. The quantity $(\rho J_0/\rho_0 J)^{\frac{1}{2}}$, which we have already considered in § 5, may be called, as in the homogeneous case, the divergence factor, amplitude factor or growth factor of the ray system. It becomes infinite when J tends to zero; hence (3.7.8) and (3.7.9) are valid only in a neighbourhood of $\mathscr{S}$ in which one, and only one, ray passes through every point. The maximal domain with this property (defined by $J>0$ and the condition that it contains $\nu=0$) is that bounded by the caustics nearest to $\mathscr{S}$. Continuation 'beyond' the caustic is not possible by means of the transport equations.

The manner in which the P_n depend on the arbitrary functions $\pi_0, \pi_1, \ldots$ can be inferred as follows. Suppose first that $\pi_0 \neq 0$ while $\pi_n=0$ for $n=1, 2, \ldots$. Then (3.7.13) can be written as

$$P_0=A\pi_0, \quad P_n=I[P_{n-1}], \tag{3.7.10}$$

where A is the divergence factor and I denotes the linear operator defined by

$$I[\psi]=-\frac{1}{2}\left(\frac{\rho}{J}\right)^{\frac{1}{2}}\int_0^\nu(\rho' J')^{\frac{1}{2}}L[\psi']\,d\nu'. \tag{3.7.11}$$

Hence we can write, in this case, $P_n=I^n[A\pi_0]$, where I^n is the nth iterate of I. It is then obvious that P_n is in general given by

$$P_n=I^n[A\pi_0]+I^{n-1}[A\pi_1]+\ldots+A\pi_n=\sum_{j=0}^{n}I^{n-j}[A\pi_j]. \tag{3.7.12}$$

8. The superposition principle

The transport equations of higher order can be used to derive series solutions of the wave equation whose leading terms represent the geometrical acoustics field. In order to obtain these, we shall first establish a superposition principle which is valid for weak solutions of the wave equation. If $P(\mathbf{x}, t)$ is a strict solution of the wave equation and $f(t)$ is a twice continuously differentiable function, then the 't-convolution'

$$\Phi = \int_{-\infty}^{\infty} P(\mathbf{x}, t')\, f(t-t')\, \mathrm{d}t' = \int_{-\infty}^{\infty} P(\mathbf{x}, t-t')\, f(t')\, \mathrm{d}t' \quad (3.8.1)$$

obviously satisfies the wave equation, provided that the twice-differentiated integrand is summable in $-\infty < t < \infty$. When P is a weak solution and f is summable but not necessarily continuous or everywhere differentiable, then it is not obvious that $L[\Phi] = 0$. However, suppose, for instance, that $f(t)$ is twice continuously differentiable and that P admits continuous derivatives up to the second order except on a characteristic $t = \tau(x)$ which carries a simple discontinuity of P. Then

$$\frac{\partial \Phi}{\partial x_\alpha} = \frac{\partial}{\partial x_\alpha}\left(\int_{-\infty}^{\tau} + \int_{\tau}^{\infty}\right) P(\mathbf{x}, t')\, f(t-t')\, \mathrm{d}t'$$

$$= -\frac{\partial \tau}{\partial x_\alpha}[P]_{\tau-0}^{\tau+0} f(t-\tau) + \int_{-\infty}^{\infty} \frac{\partial P}{\partial x_\alpha} f(t-t')\, \mathrm{d}t'.$$

When the second-order derivatives are calculated in the same manner and substituted in $L[\Phi]$ it is found that the terms arising from the discontinuity cancel because $\tau(\mathbf{x})$ satisfies the eikonal equation and $[P]_{\tau-0}^{\tau+0}$ the transport equation. This suggests that (3.8.1) satisfies the wave equation even when P is a weak solution.

A simple proof of this conjecture can be given by means of the theory of distributions. Let P be a weak solution of the wave equation and $f(t)$ an arbitrary function, and suppose that both P and f are zero for all sufficiently large negative t. Then we can form the convolution of P, considered as a distribution, and of $\delta(\mathbf{x})\, f(t)$.† This is denoted by

$$\Phi = P * \delta(\mathbf{x})\, f(t), \quad (3.8.2)$$

† The three-dimensional delta function $\delta(x_1)\,\delta(x_2)\,\delta(x_3)$ is here denoted by $\delta(\mathbf{x})$. It is defined as a distribution by the equation $\delta(\mathbf{x}) . \phi(\mathbf{x}) = \phi(0)$.

and defined by

$$\{P*\delta(\mathbf{x})\,f(t)\}\,.\,\phi(\mathbf{x},t)=P(\mathbf{x},t)\,.\,\{\delta(\mathbf{x}')\,f(t')\,.\,\phi(\mathbf{x}+\mathbf{x}',t+t')\},\tag{3.8.3}$$

where ϕ is any test-function. Thus we have

$$\begin{aligned}\{P*\delta(\mathbf{x})\,f(t)\}\,.\,\phi(\mathbf{x},t)&=\int P(\mathbf{x},t)\,\mathrm{d}\Sigma\int\delta(\mathbf{x}')\,f(t')\,\phi(\mathbf{x}+\mathbf{x}',t+t')\,\mathrm{d}\Sigma'\\&=\int P(\mathbf{x},t)\,\mathrm{d}\Sigma\int_{-\infty}^{\infty}f(t')\,\phi(\mathbf{x},t+t')\,\mathrm{d}t'\\&=\int\phi(\mathbf{x},t)\,\mathrm{d}\Sigma\int_{-\infty}^{\infty}P(\mathbf{x},t')\,f(t-t')\,\mathrm{d}t',\end{aligned}$$

whence

$$\Phi=P*\delta(\mathbf{x})\,f(t)=\int_{-\infty}^{\infty}P(\mathbf{x},t')\,f(t-t')\,\mathrm{d}t'=\int_{-\infty}^{\infty}P(\mathbf{x},t-t')\,f(t')\,\mathrm{d}t'.\tag{3.8.4}$$

The derivative of the convolution of two distributions is obtained by differentiating either factor; thus

$$L[P*\delta(\mathbf{x})\,f(t)]=\frac{1}{\rho c^2}\left[\frac{\partial^2 P}{\partial t^2}*\delta(\mathbf{x})\,f(t)\right]-\frac{1}{\rho}[\nabla^2 P*\delta(\mathbf{x})\,f(t)]+\frac{1}{\rho}\frac{\partial\rho}{\partial x_\alpha}\left[\frac{\partial P}{\partial x_\alpha}*\delta(\mathbf{x})\,f(t)\right].$$

But it is obvious from the evaluation of $P*\delta(\mathbf{x})\,f(t)$ that the coefficients on the right-hand side, which are functions of $\mathbf{x}$ only, can be taken inside the brackets, so that

$$L[\Phi]=L[P*\delta(\mathbf{x})\,f(t)]=L[P]*\delta(\mathbf{x})\,f(t).$$

By hypothesis, $L[P]=0$ in the sense of the theory of distributions, and so also $L[\Phi]=0$ in this sense. But by (3.8.4), Φ is a function, and so it is necessarily a weak solution of the wave equation. If we interpret Φ as a velocity potential, then the corresponding pressure field is given by

$$p=\frac{\partial\Phi}{\partial t}=\frac{\partial}{\partial t}\int_{-\infty}^{\infty}f(t-t')\,P(\mathbf{x},t')\,\mathrm{d}t'=\int_{-\infty}^{\infty}f(t-t')\,\mathrm{d}P(\mathbf{x},t').\tag{3.8.5}$$

A case of particular importance is that in which P is a pulse advancing into an undisturbed region of the medium, and $f(t)=0$

for $t<0$. Let $t=\tau(\mathbf{x})$ denote the front of the pulse (τ must then satisfy the eikonal equation) and put

$$\zeta=t-\tau(\mathbf{x}); \tag{3.8.6}$$

ζ is the time counted from the onset of the pulse. We also write

$$P(x,t)=\Pi(\mathbf{x},\zeta). \tag{3.8.7}$$

Now (3.8.4) becomes

$$\Phi=\int_{\tau(\mathbf{x})}^{t} P(\mathbf{x},t')\,f(t-t')\,\mathrm{d}t',$$

or, if we put $t'=\zeta'+\tau(\mathbf{x})$ and use (3.8.6) and (3.8.7),

$$\Phi=\int_0^{\zeta}\Pi(\mathbf{x},\zeta')\,f(\zeta-\zeta')\,\mathrm{d}\zeta', \tag{3.8.8}$$

with $\Phi=0$ for $\zeta<0$. By (3.8.5),

$$p=\int_0^{\zeta} f(\zeta-\zeta')\,\mathrm{d}\Pi(\mathbf{x},\zeta'), \tag{3.8.9}$$

so that one obtains $p=\Pi=P$ if one takes $f(t)=H(t)$.

It may be noted that (3.8.9) can be used to convert a solution with a simple discontinuity into one with an algebraic singularity. Let us take $f(t)=t^k$, where $k>-1$. Then

$$\Phi=\int_0^{\zeta}\Pi(\mathbf{x},\zeta')\,(\zeta-\zeta')^k\,\mathrm{d}\zeta'$$

behaves, for small positive ζ, like $P_0(\mathbf{x})\,\zeta^{k+1}/(k+1)$, where

$$P_0(\mathbf{x})=\Pi(\mathbf{x},\,+0).$$

Hence $$p\sim P_0(\mathbf{x})\,\zeta^k H(\zeta)=P_0(\mathbf{x})\,(t-\tau)^k\,H(t-\tau)$$

for small positive ζ.

9. Series expansions related to geometrical acoustics

The results obtained in the preceding sections can be combined in two different ways, one yielding a series solution valid near the front of a pulse, and the other an asymptotic series for harmonic wave trains of high frequency; in each case the leading term represents the geometrical acoustics field.

Consider first a pulse $P(\mathbf{x},t)$ propagating into a region in which the medium is undisturbed. If $t=\tau(\mathbf{x})$ is the front of the pulse then $P=0$ for $t<\tau(\mathbf{x})$, and so we have in terms of (3.7.3),

$$\left[\frac{\partial^n P}{\partial t^n}\right]_{t=\tau+0}=P_n(x), \tag{3.9.1}$$

where the P_n satisfy the transport equations. The series expansion

$$P(x,t)=\Pi(x,\zeta)=\sum_{n=0}^{\infty} P_n(x)\frac{\zeta^n}{n!}, \tag{3.9.2}$$

where ζ, defined by (3.8.6), denotes the time counted from the onset of the pulse, is then equal to P for $\zeta>0$, if it converges. This series solves the 'mixed boundary-value problem' in which $P=P_0$ on $\tau(\mathbf{x})=t$ and P is given on the surface $\mathscr{S}$ introduced in § 7. According to a theorem of Hadamard's† (who improved an earlier result due to Beudon) the series certainly converges in the neighbourhood of any point of $\mathscr{S}$ and for sufficiently small ζ if $\mathscr{S}$ is analytic and regular at this point, and the data are analytic and regular. Let us therefore assume that our series converges when ζ is sufficiently small, and substitute (3.9.2) in (3.8.8); we then obtain a more general series solution of the wave equation,

$$\Phi(\mathbf{x},t)=\sum_{n=0}^{\infty} P_n(\mathbf{x})\,f_{n+1}(\zeta), \tag{3.9.3}$$

where $f_0(\zeta)$ is an arbitrary (summable) function that vanishes for negative argument, and the f_n are obtained by repeated integration of f_0 from 0 to ζ:

$$f_{n+1}(\zeta)=\int_0^\zeta f_n(\zeta')\,d\zeta'=\int_0^\zeta f_0(\zeta')\frac{(\zeta-\zeta')^n}{n!}\,d\zeta'. \tag{3.9.4}$$

For most applications it is sufficient to assume that

$$f_0(\zeta)=\zeta^k g_0(\zeta), \tag{3.9.5}$$

where $k>-1$ and g_0 is bounded. Then

$$|f_n(\zeta)|\leqslant\frac{\sup|g_0(\zeta)|}{(n-1)!}\int_0^\zeta \zeta'^k(\zeta-\zeta')^{n-1}\,d\zeta'=\frac{k!}{(n+k)!}\zeta^{n+k}\sup_{0\leqslant\zeta'\leqslant\zeta}|g_0(\zeta')|, \tag{3.9.6}$$

so that (3.9.3) converges in the same ζ-interval as (3.9.2).

† Hadamard (1923), pp. 77–81. See also Jones (1955).

If we again interpret Φ as a velocity potential, then the pressure and the velocity field are given by the series

$$p = \sum_{n=0}^{\infty} P_n(\mathbf{x}) f_n(\zeta), \tag{3.9.7}$$

$$\rho u_\alpha = \sum_{n=0}^{\infty} \left(P_n(\mathbf{x}) \frac{\partial \tau}{\partial x_\alpha} - \frac{\partial P_{n-1}}{\partial x_\alpha} \right) f_n(\zeta) \quad (\alpha = 1, 2, 3), \tag{3.9.8}$$

(where one must put $P_{-1} = 0$). The leading terms of these series,

$$p \sim P_0(\mathbf{x}) f_0(\zeta), \quad \rho u_\alpha \sim P_0(\mathbf{x}) \frac{\partial \tau}{\partial x_\alpha} f_0(\zeta) \quad (\alpha = 1, 2, 3), \tag{3.9.9}$$

represent a 'geometrical acoustics field'; the pressure is the product of the amplitude factor $P_0(\mathbf{x})$ and the phase factor $f_0(\zeta)$, the velocity is normal to the wave fronts which are also the surfaces of constant phase, and $\rho c\,|\,\mathbf{u}\,| = p$. The other terms of the series may therefore be looked upon as supplying approximations of higher order to the first-order approximation of geometrical acoustics. The higher order terms become increasingly important as ζ, the time elapsed since the beginning of the pulse, increases, so that the series is useful for the calculation of the field immediately behind the front. It may be noted that a family of pulses of the same shape of different 'pulse lengths' is obtained if $f_0(\zeta)$ is replaced by $f_0(\zeta/\Lambda)$, where Λ is a parameter. It follows from (3.9.4) that $f_n(\zeta)$ must then be replaced by $\Lambda^n f_n(\zeta/\Lambda)$, and hence for a fixed value of ζ/Λ, p behaves like $P_0(\mathbf{x}) f_0(\zeta/\Lambda) + O(\Lambda)$.

The presence of the terms of higher order in the series implies that a pulse is distorted as it propagates, even on the linear theory. An undistorted pulse can only be obtained if, for some suitable solution P_0 of the transport equation of order zero, the P_n with $n \geq 1$ vanish; this will be the case, according to the transport equations (2.7.9), only if

$$-L[P_0] = \frac{\partial}{\partial x_\alpha}\left(\frac{1}{\rho}\frac{\partial P_0}{\partial x_\alpha}\right) = 0.$$

The consequence of this hypothesis can be investigated in the case of a homogeneous medium. It is found that it can be satisfied only if the wave fronts $\tau(\mathbf{x}) = t$ are surfaces of a certain type, the cyclides of Dupin; they include parallel planes, coaxial circular cylinders, and concentric spheres. Only in the case of plane or spherical waves can P_0 be made a single-valued function of position. Since a plane wave is a limiting case of spherical waves, the upshot is that the only single-valued undistorted pulse possible is a spherical one.†

Another, and quite different, application of the superposition principle and the transport equations is to the propagation of harmonic wave trains. Although this monograph is primarily concerned with pulses, this application is of sufficient importance to merit a short digression. The argument is due to Luneburg, who considered electromagnetic waves, and has been developed further by M. Kline (1954). The underlying idea is that a steady field (that is to say one varying sinusoidally with time) can be thought of as the asymptotic limit, as $t \to \infty$, of an unsteady field; this is common sense from a physical point of view. By means of the superposition principle, a steady field can then be linked with a certain pulse solution of the wave equation. Suppose, for instance, that a disturbance is prescribed on the fixed surface $\mathscr{S}$ to which the solutions of the transport equations have been referred, by the formula

$$p = \pi_0(\lambda, \mu)\, e^{i\omega t} H(t). \tag{3.9.10}$$

Let P be a pulse solution also defined by its values on $\mathscr{S}$,

$$P = \pi_0(\lambda, \mu)\, H(t) \tag{3.9.11}$$

with the same function π_0 as in (3.9.10). Then, if we apply (3.8.9) with $f(t) = e^{i\omega t}$ we obtain a (weak) solution of the wave equation

$$p = \int_0^{\zeta} e^{i\omega(\zeta - \zeta')}\, d\Pi(\mathbf{x}, \zeta'), \tag{3.9.12}$$

which obviously satisfies (3.9.10), and is a disturbance propagated into a region of space bounded by $\mathscr{S}$; let us denote this region by S. Now it is plausible that, as $t \to \infty$, P tends to a 'static' solution P_∞ of the wave equation in S which is a function of $\mathbf{x}$ only, satisfies

$$-L[P_\infty] = \frac{\partial}{\partial x_\alpha}\left(\frac{1}{\rho}\frac{\partial P_\infty}{\partial x_\alpha}\right) = 0,$$

reduces to π_0 on $\mathscr{S}$, and satisfies some appropriate condition at infinity, such as $P_\infty < \infty$, also, that the derivatives $\partial^n P/\partial t^n$ tend to

† These solutions of the wave equation may be called *simple progressive waves*, a concept first formulated and discussed in Courant-Hilbert (1937), pp. 448–54. The result referred to in the text is due to the author; see Friedlander (1946*b*).

zero as $t \to \infty$. This is to be expected from the physical point of view; no analytic proof of a result of this type seems to have been given. If we assume it to be true, then the Fourier-Stieltjes transform of $P(\mathbf{x}, t)$ exists and is equal to

$$U(\mathbf{x}) = \int_0^\infty e^{-i\omega\zeta} \, d\Pi(\mathbf{x}, \zeta), \tag{3.9.13}$$

and it follows from (3.9.12) that

$$\lim_{t\to\infty} (p \, e^{-i\omega t}) = U(\mathbf{x}) \, e^{-i\omega\tau(\mathbf{x})}. \tag{3.9.14}$$

Hence the unsteady field p is asymptotic to a steady field $\bar{p}$,

$$\bar{p} = U(\mathbf{x}) \, e^{i\omega[t-\tau(\mathbf{x})]}. \tag{3.9.15}$$

Let us now also assume, for simplicity, that $\Pi(\mathbf{x}, \zeta)$ admits continuous derivatives of all orders in $\zeta > 0$. Then an asymptotic expansion of $\bar{p}$ can be obtained by repeated partial integration,

$$\begin{aligned} U(\mathbf{x}, \omega) &= P_0(\mathbf{x}) + \int_0^\infty e^{-i\omega\zeta} \frac{\partial \Pi}{\delta\zeta} \, d\zeta \\ &= \sum_{n=0}^{m-1} P_n(\mathbf{x}) \, (i\omega)^{-n} + (i\omega)^{-m+1} \int_0^\infty e^{-i\omega\zeta} \frac{\partial^m \Pi}{\partial \zeta^m} \, d\zeta, \end{aligned}$$

whence

$$U(\mathbf{x}, \omega) \sim \sum_{n=0}^{\infty} P_n(\mathbf{x}) \, (i\omega)^{-n} \tag{3.9.16}$$

in the Poincaré sense, if we assume that

$$\lim_{\omega\to\infty} \left\{ \int_0^\infty e^{-i\omega\zeta} \frac{\partial^m \Pi}{\partial \zeta^m} \, d\zeta \right\} = 0,$$

which will be true, for instance, if $\partial^m \Pi / \partial \zeta^m$ is summable. This is an example of Luneburg's asymptotic series; it is the formal term-by-term Fourier-Stieltjes transform of the series expansion (3.9.2) of the pulse P, and is obviously most useful when the frequency is large. The leading term again represents the geometrical acoustics field. The most important application of these series is to the calculation of fields which result when a given wave train is reflected by an obstacle; a number of examples have recently been given by Keller, Lewis and Seckler (1956). We shall not pursue this subject here any further; the solution of pulse reflexion problems by means of series of the type (3.9.7) will be considered in chapter 4.

APPENDIX

The focusing of acoustic shocks

The mathematical problem of following a disturbance along a ray which touches a caustic can be solved by means of the explicit representation of a solution of the wave equation with given initial values. In the case of a homogeneous medium this is Poisson's solution (1.6.2)

$$\Phi(\mathbf{x}, t) = tM_{ct}[\Phi_1] + \frac{\partial}{\partial t} tM_{ct}[\Phi_0]. \tag{3.A.1}$$

Here Φ_0 and Φ_1 are the values of Φ and of $\partial\Phi/\partial t$ when $t=0$, and M_{ct} denotes the mean value over a sphere of radius ct with centre $\mathbf{x}$. This formula is valid when Φ is continuous; we therefore work with the velocity potential rather than with the pressure itself.

Let us consider a disturbance propagating into an undisturbed region, and suppose that its front at time $t=0$ is a surface $\mathscr{S}_0$. We can take the origin at a point of $\mathscr{S}_0$, the x_3-axis along the ray through this point pointing into the undisturbed region and the axes Ox_1, Ox_2 along the curvature lines of $\mathscr{S}_0$ at o. Then the equation of $\mathscr{S}_0$ is

$$x_3 = -\frac{x_1^2}{2R} - \frac{x_2^2}{2S} + \dots, \tag{3.A.2}$$

where the terms omitted are of higher order. We shall calculate Φ at $(0, 0, h)$, where $h>0$; the front reaches this point when $t=h/c$ and the singularity of Φ will be obtained by evaluating the contribution of a small neighbourhood U of the origin to the spherical means when $t-h/c$ is small. Then x_3 will be, on the sphere, of the order of $|t-h/c|$, and we can approximate Φ_0 and Φ_1 by the leading terms of (3.9.3) and (3.9.7), with $f_0(\zeta)=H(\zeta)$,

$$\Phi_0 = P_0(\mathbf{x})\left(-\frac{\nu}{c}\right)H\left(-\frac{\nu}{c}\right)+\dots, \quad \Phi_1 = P_0(\mathbf{x}).H\left(-\frac{\nu}{c}\right)+\dots,$$

where ν is the distance of the point $\mathbf{x}$ from $\mathscr{S}_0$, measured along the normal. The normal can be replaced by a parallel to the x_3-axis, to the first order, and so we have approximately

$$\left.\begin{aligned}\Phi_0 &\sim K/c\left(-\frac{x_1^2}{2R}-\frac{x_2^2}{2S}-x_3\right)H\left(-\frac{x_1^2}{2R}-\frac{x_2^2}{2S}-x_3\right),\\ \Phi_1 &\sim KH\left(-\frac{x_1^2}{2R}-\frac{x_2^2}{2S}-x_3\right),\end{aligned}\right\} \tag{3.A.3}$$

where $P_0(0)=K$.

Let us put again

$$\zeta = t - \frac{h}{c}, \tag{3.A.4}$$

and assume ζ to be small. The equation of the sphere with centre (0, 0, h) and radius ct is then

$$x_3 = -c\zeta + \frac{x_1^2 + x_2^2}{2R} + \dots,$$

and the surface element is, to the first order, $dx_1 dx_2$. Hence, by (3.A.1) and (3.A.3) the contribution from the neighbourhood of 0 is approximately

$$\Phi^* \sim \frac{K}{4\pi c^2(\zeta + h/c)} \iint dx_1 dx_2 + \frac{\partial}{\partial t}\left\{\frac{K}{4\pi c^3(\zeta + h/c)} \iint Z \, dx_1 dx_2\right\}, \tag{3.A.5}$$

where
$$Z = c\zeta - \frac{x_1^2}{2}\left(\frac{1}{R} + \frac{1}{h}\right) - \frac{x_2^2}{2}\left(\frac{1}{S} + \frac{1}{R}\right), \tag{3.A.6}$$

and the integrals are extended over that part of some convenient small domain $\mathscr{U}$ of the $x_1 x_2$-plane in which $Z \geqslant 0$. The factor $K/4\pi c^2(h/c + \zeta)$ is sensibly constant and equal to $K/4\pi hc$; also $\partial/\partial t = \partial/\partial\zeta$; hence we can replace (3.A.5) by

$$\Phi^* \sim \frac{K}{2\pi hc} \iint dx_1 dx_2, \quad p^* \sim \frac{\partial \Phi^*}{\partial \zeta}. \tag{3.A.7}$$

Let us first consider the case where $R < 0$, $S < 0$, so that $\mathscr{S}_0$ is concave towards the undisturbed region at 0. We may suppose that

$$0 < -R < -S. \tag{3.A.8}$$

We must then distinguish three principal cases,

$$\text{(i)}\ 0 < h < -R; \quad \text{(ii)}\ -R < h < -S; \quad \text{(iii)}\ -S < h;$$

there are also two transitional cases, $h = -R$ and $h = -S$.

In case (i), $R + h < 0$, $S + h < 0$, and so by (3.A.6) the inequality $Z \geqslant 0$ is not satisfied anywhere when $\zeta < 0$, and defines an ellipse with semi-axes $\{2Rhc\zeta/(R+h)\}^{\frac{1}{2}}$, $\{2Sc\zeta h/(S+h)\}^{\frac{1}{2}}$ when $\zeta > 0$. Hence

$$\Phi^* = 0 \quad \left(t < \frac{h}{c}\right); \qquad \Phi^* \sim K\left\{\frac{RS}{(R+h)(S+h)}\right\}^{\frac{1}{2}} \left(t - \frac{h}{c}\right), \left(t > \frac{h}{c}\right),$$

so that the pressure has a discontinuity

$$K\left\{\frac{RS}{(R+h)(S+h)}\right\}^{\frac{1}{2}} \tag{3.A.9}$$

at time $t = h/c$; this agrees with the value derived from the transport equation, the divergence factor being given by (3.6.7).

In case (ii), we have $R + h > 0$, $S + h < 0$. Hence $Z = 0$ is a hyperbola; the domains of integration are different for $\zeta < 0$ and $\zeta > 0$, and are as shown in fig. 3.3. The requisite restriction to a neighbourhood of the origin can be effected by adding the condition $|x_1| \leqslant \epsilon$, where ϵ is small. Hence (3.A.7) gives

$$p^* \sim \frac{K}{\pi h}\left(\frac{2Sh}{S+h}\right)^{\frac{1}{2}} \int \left(\frac{(R+h)\,x_1^2}{-2Rh} + c\zeta\right)^{-\frac{1}{2}} dx_1,$$

where the integral is over $(0, \epsilon)$ when $\zeta > 0$ and over

$$\{2Rhc\zeta/(R+h)\}^{\frac{1}{2}} < x_1 < \epsilon$$

when $\zeta < 0$. Hence

$$p^* \sim \frac{2K}{\pi}\left(\frac{RS}{(R+h)\,|S+h|}\right)^{\frac{1}{2}} \cosh^{-1}\left\{\epsilon\left(\frac{R+h}{2Rhc\zeta}\right)^{\frac{1}{2}}\right\} \quad (\zeta < 0);$$

$$p^* \sim \frac{2K}{\pi}\left(\frac{RS}{(R+h)\,|S+h|}\right)^{\frac{1}{2}} \sinh^{-1}\left\{\epsilon\left(\frac{R+h}{2|R|hc\zeta}\right)^{\frac{1}{2}}\right\} \quad (\zeta > 0),$$

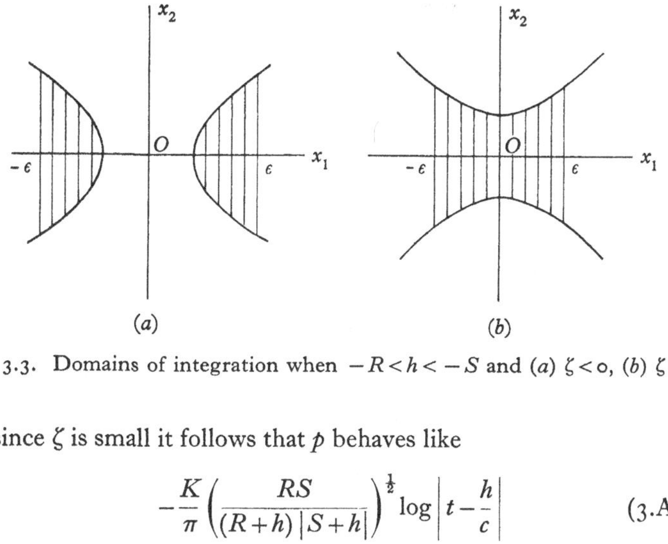

Fig. 3.3. Domains of integration when $-R < h < -S$ and (a) $\zeta < 0$, (b) $\zeta > 0$.

and since ζ is small it follows that p behaves like

$$-\frac{K}{\pi}\left(\frac{RS}{(R+h)\,|S+h|}\right)^{\frac{1}{2}} \log\left|t - \frac{h}{c}\right| \qquad (3.A.10)$$

augmented by a term which remains finite at $t = h/c$, when $|t - h/c|$ is small.

In case (iii), the integration is over the exterior of an ellipse with semi-axes $\{2Rhc\zeta/(R+h)\}^{\frac{1}{2}}$, $\{2Shc\zeta/(S+h)\}^{\frac{1}{2}}$ for $\zeta < 0$, and over the whole of the small neighbourhood $\mathscr{U}$ for $\zeta > 0$. Hence if we denote the area of $\mathscr{U}$ by δ we find that

$$\Phi^* \sim \frac{K}{2\pi hc}\left\{\delta + \frac{2\pi hc\zeta(RS)^{\frac{1}{2}}}{\{(R+h)(S+h)\}^{\frac{1}{2}}}\right\} \quad (\zeta < 0); \quad \Phi^* \sim \frac{K\delta}{2\pi hc} \quad (\zeta > 0),$$

so that p changes discontinuously by the amount

$$-K\left\{\frac{RS}{(R+h)(S+h)}\right\}^{\frac{1}{2}} \tag{3.A.11}$$

at time $t=h/c$.

The transitional cases can be dealt with similarly, but one must take terms of order higher than the second in the equation of $\mathscr{S}_0$ into account. If we replace (2) by

$$x_3 = -\frac{x_1^2}{2R} - \frac{x_2^2}{2S} + Bx_1^3 + Cx_1^2x_2 + Dx_1x_2^2 + Ex_2^3 + \ldots$$

and take account of the higher order terms in the equation of the sphere, then we find that the integration is, for instance, in the case $h=-R$ to be confined to

$$c\zeta \geqslant x_1^2\frac{R-S}{2RS} - \frac{x_1^2+x_2^2}{2R^2}c\zeta - Bx_1^3 - Cx_1^2x_2 - Dx_1x_2^2 - Ex_2^3 + \ldots$$

and to a small neighbourhood of the origin. It is easily seen that for small ζ, x_1 will be of order $|\zeta|^{\frac{1}{2}}$ and x_2 of order $|\zeta|^{\frac{1}{3}}$, so that when only the significant terms are retained the inequality becomes

$$c\zeta \geqslant x_1^2\frac{R-S}{2RS} - Ex_2^3.$$

One can then show that

$$p^* \sim \frac{K}{2\pi|R|}\left(\frac{2RS}{R-S}\right)^{\frac{1}{2}} E^{-\frac{1}{3}}|ct-h|^{-\frac{1}{6}}\int_1^{\infty}(\xi^3-1)^{-\frac{1}{2}}\,d\xi \quad \left(t<\frac{h}{c}\right),$$

$$p^* \sim \frac{K}{2\pi|R|}\left(\frac{2RS}{R-S}\right)^{\frac{1}{2}} E^{-\frac{1}{3}}|ct-h|^{-\frac{1}{6}}\int_{-1}^{\infty}(\xi^3+1)^{-\frac{1}{2}}\,d\xi \quad \left(t>\frac{h}{c}\right).$$

Similar results are obtained for $h=-S$. If $E=0$, fourth-order terms must be taken into account, and a different power law is obtained.

When the front $\mathscr{S}_0$ is anticlastic at 0, say $R<0$, $S>0$, then there is only one focal point on the ray in the undisturbed region, $x_3=-R$. Now for $h>-R$, $(R+h)/Rh$ and $(S+h)/Sh$ have the same signs as in the case (ii) considered above; hence we shall obtain the same result as in that case, namely, (3.A.10).†

† For an investigation of the analogous problem in the harmonic case see Keller and Kay (1954).

CHAPTER 4

THE APPLICATION OF GEOMETRICAL ACOUSTICS TO REFLEXION PROBLEMS

1. Introduction

In the last two chapters we have developed some aspects of the theory of the wave equation and discussed their interpretation in terms of geometrical acoustics. We now turn to the solution of specific reflexion and diffraction problems. The distinction between diffracted and reflected fronts was drawn in chapter 1 (§§ 1.5 and 1.6). Diffracted fronts propagate into a shadow and are normal to rays which touch the reflector; reflected fronts are normal to reflected rays issuing from the points of the reflector that are accessible in physical space along incident rays. The present chapter deals with the calculation of a reflected pulse at and immediately behind its front; some explicit solutions of reflexion problems are also discussed. Diffraction will be treated in chapters 5 and 6. The analytical behaviour of the secondary field at a diffracted front is different from that at a reflected front, and diffraction problems are in general more difficult. But in the case of a reflected front the theory of chapter 3 can be used. In principle any reflected pressure pulse can be expanded as a series of the type (3.9.7) immediately behind its front so that a very general method for the solution of reflexion problems is available. In practice the requisite calculations are usually so heavy that it is difficult to calculate more than say the first two terms of the series and very few examples have been worked out in detail. A comprehensive account of such solutions has been given by Keller, Lewis and Seckler (1956) for periodic incident fields, using Luneburg's asymptotic series.

The series expansion of a reflected pulse is obtained as follows. Let $t = \tau_i(\mathbf{x})$ be the front of the incident field and suppose that this field is already given in the form (3.9.7) and (3.9.8)

$$\left.\begin{aligned} p_i &= \sum_{s=0}^{\infty} P_s^{(i)}(\mathbf{x})\, f_s(\zeta_i), \\ \rho\mathbf{u}_i &= \sum_{s=0}^{\infty} \{P_s^{(i)} \operatorname{grad} \tau_i - \operatorname{grad} P_{s-1}^{(i)}\}\, f_s(\zeta_i), \quad P_{-1}^{(i)} = 0, \end{aligned}\right\} \qquad (4.1.1)$$

where
$$\zeta_i = t - \tau_i(\mathbf{x}), \tag{4.1.2}$$
and the f_s are the repeated integrals of f_0. The reflected front can then be calculated by means of the reflexion law of geometrical optics. We write its equation as $t = \tau_r(\mathbf{x})$ and note that $\tau_i(\mathbf{x}) = \tau_r(\mathbf{x})$ holds on the reflector $\mathscr{R}$. We now assume a similar series expansion for the reflected pulse,

$$\left.\begin{aligned} p_r &= \sum_{s=0}^{\infty} P_s^{(r)} f_s(\zeta_r), \\ \rho\mathbf{u}_r &= \sum_{s=0}^{\infty} \{P_s^{(r)} \operatorname{grad} \tau_r - \operatorname{grad} P_{s-1}^{(r)}\} f_s(\zeta_r), \quad P_{-1}^{(r)} = 0, \\ \zeta_r &= t - \tau_r(\mathbf{x}). \end{aligned}\right\} \tag{4.1.3}$$

The coefficients $P_s^{(r)}$ satisfy the transport equations
$$\operatorname{Tr}(\tau_r, P_s^{(r)}) = -L(P_{s-1}^{(r)}) \quad (s = 0, 1, 2, \ldots). \tag{4.1.4}$$
Hence they can be determined by integration along the reflected rays if their values on $\mathscr{R}$ are known. But these can be deduced from the boundary condition which holds on $\mathscr{R}$. Let us suppose for simplicity that the reflector is rigid and fixed. Then
$$(\mathbf{u}_i + \mathbf{u}_r) . \mathbf{n} = 0, \tag{4.1.5}$$
where $\mathbf{n}$ is the normal to $\mathscr{R}$ drawn (say) into physical space. Since $\zeta_i = \zeta_r$ on $\mathscr{R}$ it follows from (4.1.1) and (4.1.3) that
$$P_s^{(i)} \frac{\partial \tau_i}{\partial n} - \frac{\partial P_{s-1}^{(i)}}{\partial n} + P_s^{(r)} \frac{\partial \tau_r}{\partial n} - \frac{\partial P_{s-1}^{(r)}}{\partial n} = 0 \quad (s = 0, 1, 2, \ldots),$$
on $\mathscr{R}$. The incident and reflected rays are equally inclined to $\mathbf{n}$ so that
$$\frac{\partial \tau_i}{\partial n} = -c \cos \gamma, \quad \frac{\partial \tau_r}{\partial n} = c \cos \gamma,$$
where γ is the angle between the incident ray and $-\mathbf{n}$. Hence
$$P_0^{(r)} = P_0^{(i)}, \quad P_s^{(r)} = P_s^{(i)} + \frac{1}{c \cos \gamma} \left(\frac{\partial P_{s-1}^{(r)}}{\partial n} - \frac{\partial P_{s-1}^{(i)}}{\partial n} \right) \quad (s = 1, 2, \ldots), \tag{4.1.6}$$
holds on $\mathscr{R}$. These recurrence equations and the transport equations (4.1.4) determine the $P_s^{(r)}$. For, assuming that the $P_s^{(r)}$ for $s = 0, 1, \ldots, m-1$ have already been determined, we see that (4.1.6) gives the

value of P_m on $\mathscr{R}$ and hence P_m can be found by integrating the appropriate transport equation. The method can of course be applied only to the domain of direct reflexion formed by the points that can be reached along reflected rays; the series is valid only at and near the 'true' front corresponding to the earliest arrival of the reflected pulse and breaks down near a caustic or a shadow boundary.

It has already been stated that the actual calculation of the P_s usually requires very heavy algebra. It is therefore best to divide the exposition of the method into two stages by beginning with a discussion of the leading term $P_0^{(r)} f_0(\zeta_r)$ which is the geometrical acoustics approximation to the reflected pulse. If the incident pulse begins with an acoustic shock we can take $f_0(\zeta) = H(\zeta)$. Then $P_0^{(i)}$ and $P_0^{(r)}$ are strengths of the incident and reflected acoustic shocks respectively. (This is also true for the reflexion of acoustic shocks which are not pulse fronts.) The calculation of P_0 involves the determination of the reflected ray system and of its divergence factor. When the medium is homogeneous the results derived in § 3.6 can be used for this purpose. Let $\mathbf{x}$ be a point on a reflected ray which issues from the point $\mathbf{x}_0$ of $\mathscr{R}$. Let also σ be the distance of $\mathbf{x}_0$ from $\mathbf{x}$ and F_1, F_2 be the principal radii of curvature of the reflected front at $\mathbf{x}_0$. Then by (3.6.7) and (4.1.6)

$$P_0^{(r)}(\mathbf{x}) = A P_0^{(i)}(\mathbf{x}_0), \tag{4.1.7}$$

where

$$A = \left\{\frac{F_1 F_2}{(F_1+\sigma)(F_2+\sigma)}\right\}^{\frac{1}{2}} \tag{4.1.8}$$

is the divergence factor of the reflected rays normalized so that its value on $\mathscr{R}$ is unity. On the reflector itself, the initial pressure rise is doubled; this is a consequence of (4.1.6) and is true in any medium.

We shall suppose from now on that the medium is homogeneous, and proceed to evaluate the divergence factor A when the incident front is plane or spherical; these are of course the most important cases in practice in the case of a homogeneous medium.

2. Reflexion of a plane pulse

The simplest case is that of the reflexion of a plane pulse by a cylindrical reflector whose generators are parallel to the incident fronts. The reflected pulse is then two-dimensional. Denoting the coordinates now by x, y and z, we take the z-axis parallel to the generators and the x-axis opposite to the direction of incidence. Then the equation of the incident fronts may be taken to be $ct+x=0$. The reflector $\mathscr{R}$ corresponds to a curve in the xy-plane which we also denote by $\mathscr{R}$. The calculations are simplified considerably if we assume $\mathscr{R}$ to be given by the parametric equations

$$x = x_0(\psi), \quad y = y_0(\psi), \tag{4.2.1}$$

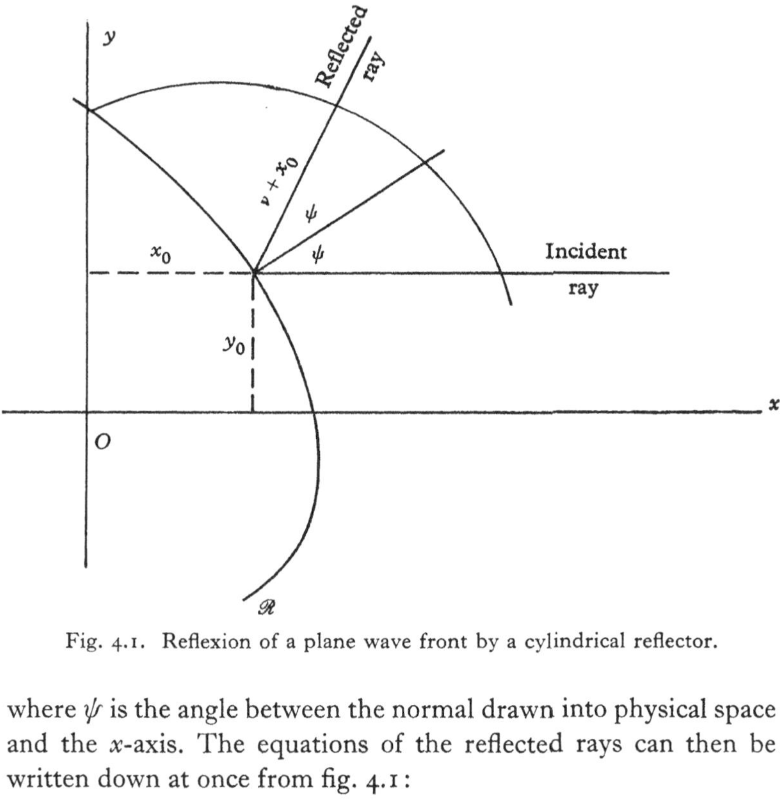

Fig. 4.1. Reflexion of a plane wave front by a cylindrical reflector.

where ψ is the angle between the normal drawn into physical space and the x-axis. The equations of the reflected rays can then be written down at once from fig. 4.1:

$$\left.\begin{aligned} x &= x_0(\psi) + \{\nu + x_0(\psi)\}\cos 2\psi, \\ y &= y_0(\psi) + \{\nu + x_0(\psi)\}\sin 2\psi. \end{aligned}\right\} \tag{4.2.2}$$

Here the parameter ν has been chosen so that the equation of the reflected front at time t is $ct=\nu$; it is related to the distance σ defined previously by the equation $\sigma = \nu + x_0(\psi)$. Since one of the principal radii of curvature of the reflected fronts is in the present case infinite, the formula (4.1.8) becomes

$$A=\left(\frac{F}{\nu+x_0+F}\right)^{\frac{1}{2}}. \tag{4.2.3}$$

To calculate F we note that the angle between adjacent reflected rays is $2\delta\psi$, while the cross-section of the ray-tube at (x_0, y_0) is $\delta s\cos\psi$, δs being the element of arc on $\mathscr{R}$. Thus $2F\delta\psi=\delta s\cos\psi$. But $\delta s=f\delta\psi$, where f is the radius of curvature of $\mathscr{R}$ at the point $x_0(\psi)$, $y_0(\psi)$. Hence

$$F=\tfrac{1}{2}f\cos\psi \tag{4.2.4}$$

and

$$A=\left(\frac{f\cos\psi}{f\cos\psi+2x_0+2\nu}\right)^{\frac{1}{2}}. \tag{4.2.5}$$

It is obvious that the condition that the foot of every reflected ray must be accessible in physical space along an incident ray implies in the present case that $|\psi|\leqslant\frac{1}{2}\pi$. The limiting values $\frac{1}{2}\pi$, $-\frac{1}{2}\pi$ can be attained in two ways. If $\mathscr{R}$ is limited in the positive or negative y-direction then some of the incident rays do not meet it at all, and there will be 'glancing' incident rays which touch $\mathscr{R}$ at the points where either $\psi=\frac{1}{2}\pi$ or $\psi=-\frac{1}{2}\pi$. The corresponding reflected rays are the prolongations of the incident rays; they are part of the boundary of the shadow which is formed in this case. The formula (4.2.5) gives $A=0$ on the shadow boundary. Secondly, $\mathscr{R}$ may have a point of inflexion at which the tangent is parallel to the x-axis; in this case the reflected ray coincides with the incident ray, and again $A=0$. In both cases A is discontinuous at the reflector where $A=1$ by definition. When $|\psi|<\frac{1}{2}\pi$ then F has the same sign as f; hence the reflected wave front is convex or concave respectively when $\mathscr{R}$ is convex or concave respectively towards $x>0$ at the foot of the incident ray as long as $F/(\nu+x_0+F)>0$, that is to say, when there is no focusing. There is a focal point on any reflected ray that issues from a point where $\mathscr{R}$ is concave; if $\mathscr{R}$ has a point of inflexion then that point is the focal point on the corresponding reflected ray.

If the reflector is everywhere convex towards $x>0$ then $f>0$. The reflected wave fronts expand without focusing and geometrical acoustics can be applied unequivocally at all points of the domain of direct reflexion. It is then easy to calculate the 'far-field' of a reflected acoustic shock. When ν is large then by (4.2.1)

$$r \simeq \nu + x_0(\psi), \quad \theta \simeq \tfrac{1}{2}\psi, \tag{4.2.6}$$

where (r, θ) are polar coordinates, and

$$A \simeq \left\{\frac{f(\frac{1}{2}\theta)\cos\frac{1}{2}\theta)}{2r}\right\}^{\frac{1}{2}}. \tag{4.2.7}$$

Thus the ratio of the energy scattered per unit solid angle in the direction θ to the energy incident per unit area of the incident wave front is $\frac{1}{2}f(\frac{1}{2}\theta)\cos\frac{1}{2}\theta$.

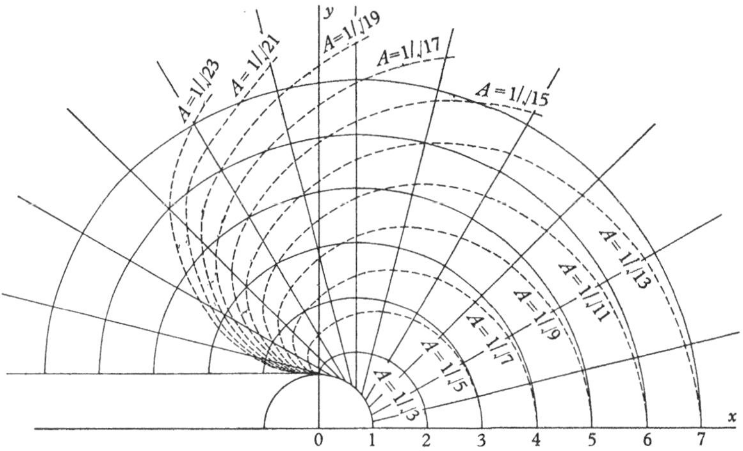

Fig. 4.2. Reflexion of a plane acoustic shock wave by a circular cylinder. ——, reflected fronts; ----, curves of constant pressure rise A at the reflected front. The curve $A = 1/\sqrt{3}$ coincides with the reflected front $\nu = 0$.

The information furnished by geometrical acoustics can be exhibited by drawing wave fronts $\nu = \text{const.}$ and curves of constant amplitude $A = \text{const.}$ Two examples of such curves are shown in figs. 4.2 and 4.3. Fig. 4.2 shows the curves when $\mathscr{R}$ is a circular cylinder. Then f is constant, say $f = a$, and (4.2.1) and (4.2.5) become

$$\left.\begin{aligned} x &= a\cos\psi + (\nu + a\cos\psi)\cos 2\psi, \\ y &= a\sin\psi + (\nu + a\cos\psi)\sin 2\psi, \\ A &= \left(\frac{a\cos\psi}{2\nu + 3a\cos\psi}\right)^{\frac{1}{2}} \quad (|\psi| \leqslant \tfrac{1}{2}\pi). \end{aligned}\right\} \tag{4.2.8}$$

As there is symmetry about the x-axis, the figure shows only the curves in $y \geqslant 0$; the diffracted fronts have been omitted. The fronts shown correspond to integral values of ν/a. The curves of constant A selected meet the x-axis at $x = (n+1)a$ $(n = 1, 2, \ldots)$; the value of A along such a curve is $1/\surd(2n+1)$ (the curve $A = 1/\surd 3$ coincides with the front $\nu = 0$). The figure shows that most of the energy is scattered into $x > 0$. In fact, the far-field (4.2.7) is

$$\left(\frac{a}{2r}\cos\tfrac{1}{2}\theta\right)^{\frac{1}{2}}.$$

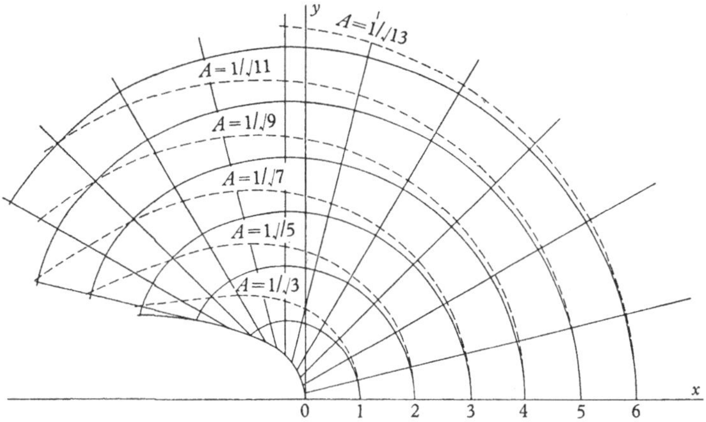

Fig. 4.3. Reflexion of a plane acoustic shock wave by the cylinder $x = \log\cos y$. ——, reflected fronts; ----, curves of constant pressure rise A at the reflected front.

The total scattered energy is, according to the intensity law, equal to the energy incident on the cross-section of the cylinder. The proportion of energy scattered into $x > 0$ is therefore

$$\frac{1}{2a}\int_{-\frac{1}{2}\pi}^{\frac{1}{2}\pi} A^2 r\,\mathrm{d}\theta = \frac{1}{\surd 2}.$$

Fig. 4.3 shows the wave fronts and curves of constant amplitude A for a reflector with the equations

$$x_0 = a\log\cos\psi, \quad y_0 = a\psi, \quad |\psi| \leqslant \tfrac{1}{2}\pi. \tag{4.2.9}$$

In this case $f = a\sec\psi$ so that

$$A = \left(\frac{a}{a + 2a\log\cos\psi + 2\nu}\right)^{\frac{1}{2}}. \tag{4.2.10}$$

The far-field is independent of θ. The curve $\mathscr{R}$ is situated in $x \leqslant 0$; it is symmetrical about the x-axis, and as $x \to \infty$, y tends either to $\frac{1}{2}a\pi$ or to $-\frac{1}{2}a\pi$, so that $\mathscr{R}$ is asymptotically of constant width $a\pi$.†

The reflexion of a plane pulse by a surface of revolution can be discussed similarly. We take the axis of symmetry as the x-axis and the incident front as $ct+x=0$. If r now denotes the distance from the x-axis, then the reflector $\mathscr{R}$ can be taken to have the equations $x=x_0(\psi)$, $r=r_0(\psi)$, where ψ is the angle between the normal drawn into physical space and the x-axis. These rays are in meridional planes and their equations are obtained from (4.2.2) when y, y_0 are replaced by r, r_0 respectively:

$$\left.\begin{aligned} x &= x_0(\psi)+(x_0(\psi)+\nu)\cos 2\psi, \\ r &= r_0(\psi)+(x_0(\psi)+\nu)\sin 2\psi. \end{aligned}\right\} \tag{4.2.11}$$

The divergence factor is given by (4.1.8) with $\sigma=\nu+x_0$ and $F_1=F=\frac{1}{2}f\cos\psi$ as before; the wave fronts are surfaces of revolution about the x-axis so that $F_2=r/\sin 2\psi$, the distance from the axis measured along a reflected ray. Hence

$$A=\left(\frac{r_0}{r}\frac{F}{F+\nu+x_0}\right)^{\frac{1}{2}}=\left\{\frac{fr_0\cos\psi}{(r_0+(\nu+x_0)\sin 2\psi)(f\cos\psi+2x_0+2\nu)}\right\}^{\frac{1}{2}}. \tag{4.2.12}$$

In addition to focusing in meridional planes, reflected rays may also meet on the x-axis. When ν is large then $\nu+x_0 \simeq R$ and $2\psi \simeq \theta$, where R is the distance from the origin and θ is the co-latitude $\cos^{-1}(x/R)$. Hence

$$A \simeq \frac{1}{2R}\{f(\tfrac{1}{2}\theta)\,r_0(\tfrac{1}{2}\theta)\operatorname{cosec}\tfrac{1}{2}\theta\}^{\frac{1}{2}}. \tag{4.2.13}$$

Fig. 4.4 shows the wave fronts $\nu=$ const. and the curves of constant amplitude A in the $x\,r$-plane when $\mathscr{R}$ is a sphere $R=a$. The wave fronts shown correspond to integral values of ν/a, and the curves of constant A meet the x-axis at $x=(n+1)a$ $(n=1, 2, \ldots)$; along such a curve $A=1/(2n+1)$. It will be observed that in this case (4.2.13) shows that the 'far-field' is independent of θ.

† The potential flow past this curve which is a uniform flow for large positive x can be obtained easily by combining a source at the origin and a uniform flow-field.

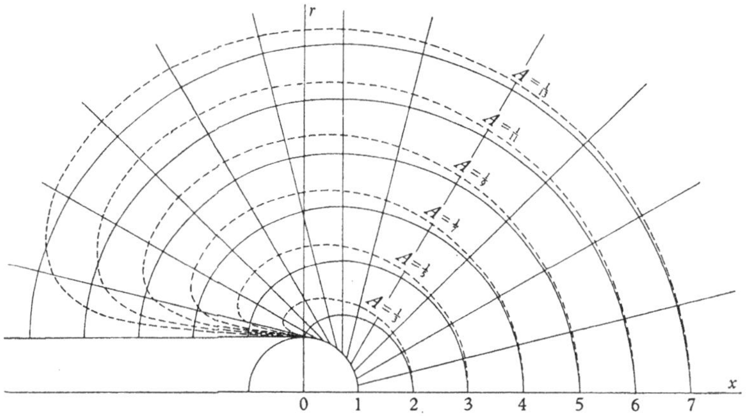

Fig. 4.4. Reflexion of a plane acoustic shock wave by a sphere. ——, reflected fronts; ----, curves of constant pressure rise A at the reflected front.

3. Reflexion of a spherical pulse by a surface of revolution

Another relatively simple example is the reflexion of a spherical pulse by a surface of revolution, provided that the source is on the axis of symmetry. Let (x, r) again be cylindrical polar coordinates, and let the equation of the reflector be again $x = x_0(\psi)$, $r = r_0(\psi)$, with the same meaning of ψ as in the last section. Then (if we assume that physical space contains all points with sufficiently large x)

$$x_0'(\psi) = -f \sin \psi, \quad r_0'(\psi) = f \cos \psi, \tag{4.3.1}$$

where f is the radius of curvature of the reflector $\mathscr{R}$ in a meridional plane. Let the source be at $x = d$, $r = 0$, and let R_0 be the distance of a point (x_0, r_0) of the reflector from the source and γ the angle between the incident ray and the normal (fig. 4.5). Then

$$R_0 \sin(\gamma - \psi) = r_0, \quad R_0 \cos(\gamma - \psi) = d - x_0. \tag{4.3.2}$$

From these equations and (4.3.1) it follows that

$$\frac{dR_0}{d\psi} = f \sin \psi, \quad \frac{d\gamma}{d\psi} = 1 + \frac{f \cos \gamma}{R_0}. \tag{4.3.3}$$

The equations of the reflected fronts and rays are

$$\left.\begin{aligned} x &= x_0(\psi) + (\nu - R_0)\cos(\gamma + \psi), \\ r &= r_0(\psi) + (\nu - R_0)\sin(\gamma + \psi). \end{aligned}\right\} \tag{4.3.4}$$

The divergence factor is

$$A=\left(\frac{F}{F+\nu-R_0}\frac{r_0}{r}\right)^{\frac{1}{2}}, \qquad (4.3.5)$$

where F again denotes the radius of curvature of the reflected front at (x_0, r_0) in a meridional plane. Now the inclination of a reflected ray to the x-axis changes by $\delta(\gamma+\psi)$ when ψ is increased to $\psi+\delta\psi$, and the normal distance between adjacent reflected rays is $\delta s \cos\gamma$ at $\mathscr{R}$; hence

$$F\delta(\gamma+\psi)=\delta s\cos\gamma=f\cos\gamma\,\delta\psi,$$

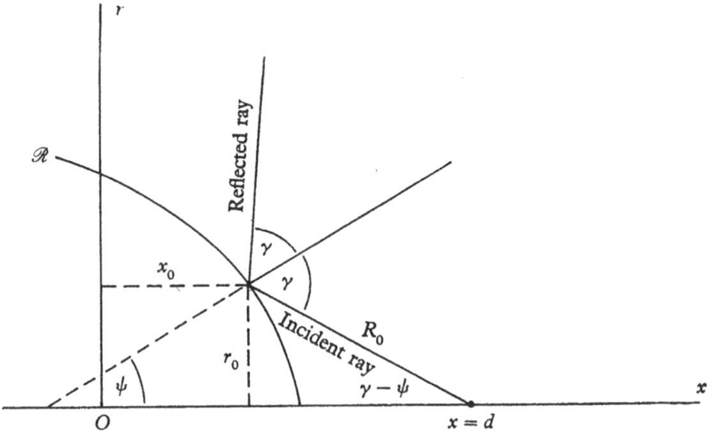

Fig. 4.5. Reflexion of a spherical wave front by a surface of revolution.

and so by (4.3.2)

$$F=\frac{R_0 f\cos\gamma}{2R_0+f\cos\gamma}. \qquad (4.3.6)$$

Substituting this and the value of r from (4.3.4) in (4.3.5) we find

$$A=\left\{\left[1+\left(\frac{2}{f\cos\gamma}+\frac{1}{R_0}\right)(\nu-R_0)\right]\left[1+\frac{\sin(\gamma+\psi)}{r_0}(\nu-R_0)\right]\right\}^{-\frac{1}{2}} \qquad (4.3.7)$$

for the divergence factor. To obtain the strength of the reflected acoustic shock we must multiply by the strength of the incident shock. Since the incident pulse is a spherical one we may take $P_0^{(i)}=1/R$, where R is the distance from the source point; hence

$$P_0^{(r)}=\frac{A}{R_0}. \qquad (4.3.8)$$

To obtain the value of P_0 on the axis we note that for small ψ

$$r_0 \simeq f\psi, \quad \sin(\gamma+\psi) \simeq \left(2+\frac{f}{R_0}\right)\psi$$

by (4.3.1) and (4.3.3). Thus since $\nu - R_0 = x - x_0$ for $\psi = 0$,

$$[P_0^{(r)}]_{\psi=0} = \frac{f}{(d+x-2x_0)f + 2(d-x_0)(x-x_0)}, \tag{4.3.9}$$

where f and x_0 must be evaluated for $\psi = 0$. The symmetry of this expression with respect to $d - x_0$ and $x - x_0$ is an instance of a general reciprocity theorem; the amplitude P_0 at $\mathbf{x}_2$ due to a source at a point $\mathbf{x}_1$ is equal to the amplitude at $\mathbf{x}_1$ due to a similar source at $\mathbf{x}_2$.

The formula (4.3.9) can be interpreted as follows. Let us suppose for simplicity that $x_0 = 0$ so that d is the distance of the source from the obstacle measured along the axis. Rays adjacent to the axis are then focused according to elementary ('second order') geometrical optics at a point $x = d'$, where

$$\frac{2}{f} + \frac{1}{d} + \frac{1}{d'} = 0. \tag{4.3.10}$$

We can therefore write (4.3.9) in the form

$$[P_0^{(r)}]_{\psi=0} = -\frac{d'}{d}\frac{1}{x-d'}. \tag{4.3.11}$$

Thus the reflected pulse on the axis is in the geometrical acoustics approximation equal to a spherical pulse originating at the optical image $x = d'$ at time $(d+d')/c$ which differs from the incident pulse only by the factor $-(d'/d)$. On any other ray there are in general two foci, one where the ray meets the axis and the other given by

$$\frac{1}{\nu - R_0} + \frac{1}{R_0} + \frac{2}{f\cos\gamma} = 0. \tag{4.3.12}$$

This is the relation (4.3.11) with f replaced by an 'apparent' focal length $f\cos\gamma$. Either or both the foci may of course be virtual images, that is to say, be on the reflected ray prolonged out of physical space.

The corresponding two-dimensional problem is the reflexion of a cylindrical pulse due to a line source by a cylinder whose generators are parallel to the line source. If we take the source at $x=d$, $y=0$ and denote the distance from it by r, then the geometrical optics approximation to the incident pulse is

$$r^{-\frac{1}{2}} f_0\left(t-\frac{r}{c}\right), \tag{4.3.13}$$

since $r^{-\frac{1}{2}}$ satisfies the appropriate transport equation. The reflected pulse is then to the same order of approximation

$$P_0^{(r)} f_0\left(t-\frac{\nu}{c}\right).$$

The equations for the reflected rays and fronts analogous to (4.3.4) are

$$\left.\begin{aligned} x &= x_0+(\nu-r_0)\cos(\psi+\gamma), \\ y &= y_0+(\nu-r_0)\sin(\psi+\gamma), \end{aligned}\right\} \tag{4.3.14}$$

where $x_0(\psi)$, $y_0(\psi)$ are the coordinates of a point of the reflector as functions of the inclination ψ of the normal to the x-axis, ψ is the angle between the normal and the incident ray and r_0 the distance of (x_0, y_0) from the source. Then

$$P_0^{(r)} = \frac{1}{r_0^{\frac{1}{2}}}\left(\frac{F}{F+\nu-r_0}\right)^{\frac{1}{2}},$$

where F is again the radius of curvature of the reflected front at (x_0, y_0); this is obviously given by (4.3.6) with R_0 replaced by r_0, so that

$$P_0^{(r)} = \left\{\frac{f\cos\gamma}{r_0 f\cos\gamma+(2r_0+f\cos\gamma)(\nu-r_0)}\right\}^{\frac{1}{2}}. \tag{4.3.15}$$

The strength of the shock due to the reflexion of a spherical shock by any surface can also be calculated. The resulting formula is rather a complicated one; it is derived in the appendix to this chapter.

4. Series expansion of a reflected pulse

The practical difficulties encountered in the calculation of the series (4.1.3) can be illustrated by considering a simple example. Let a plane pressure pulse

$$p_r = f_0(ct+x) \tag{4.4.1}$$

be incident in a homogeneous medium on a fixed and rigid reflector whose boundary $\mathscr{R}$ is a cylinder parallel to the z-axis. This is the case considered in §4.2. Dropping the superscripts we can write (4.1.3) in the slightly modified form

$$p_r = \sum_{s=0}^{\infty} P_s(x, y)\, f_s(ct - \nu), \qquad (4.4.2)$$

where ν has the same meaning as in §4.2. It is easily shown that the transport equations are then

$$2\frac{\partial P_0}{\partial \nu} + P_0 \nabla^2 \nu = 0, \qquad (4.4.3)$$

$$2\frac{\partial P_s}{\partial r} + P_s \nabla^2 \nu = \nabla^2 P_{s-1} \quad (s = 1, 2, \ldots). \qquad (4.4.4)$$

In the notation of §4.1 we have $P_0^{(i)} = 1$, $P_s^{(i)} = 0$ $(s \geqslant 1)$, and $\gamma = \psi$, where ψ is the angle between the normal to $\mathscr{R}$ and the x-axis. Since the argument of the f_s is $ct - \nu$ and not $t - \nu/c$ as in §4.1 the boundary conditions (4.1.6) are therefore in the present case

$$[P_0] = 1, \qquad (4.4.5)$$

$$[P_s] = \left[\frac{\partial P_{s-1}}{\partial n}\right] \sec\psi \quad (s \geqslant 1), \qquad (4.4.6)$$

where the square brackets indicate that the quantities enclosed by them are to be evaluated on $\mathscr{R}$.

We can now proceed by introducing ψ and ν as curvilinear co-ordinates. Since $x_0' = -f\sin\psi$, $y_0' = f\cos\psi$, it follows from (4.2.2) that

$$\left.\begin{aligned} dx &= -h\sin 2\psi\, d\psi + d\nu\cos 2\psi, \\ dy &= h\cos 2\psi\, d\psi + d\nu\sin 2\psi, \end{aligned}\right\} \qquad (4.4.7)$$

where

$$h = 2(\nu + x_0) + f\cos\psi. \qquad (4.4.8)$$

The transport equations (4.4.3) and (4.4.4) therefore become

$$2\frac{\partial P_0}{\partial \nu} + \frac{2}{h} P_0 = 0, \qquad (4.4.9)$$

$$2\frac{\partial P_s}{\partial \nu} + \frac{2}{h} P_s = \frac{1}{h}\frac{\partial}{\partial \psi}\left(\frac{1}{h}\frac{\partial P_{s-1}}{\partial \psi}\right) + \frac{1}{h}\frac{\partial}{\partial \nu}\left(h\frac{\partial P_{s-1}}{\partial \nu}\right) \quad (s = 1, 2, \ldots). \qquad (4.4.10)$$

We must also transform the boundary conditions (4.4.6). Clearly

$$\left[\frac{\partial P_{s-1}}{\partial n}\right]=\left[\cos\psi\frac{\partial P_{s-1}}{\partial x}+\sin\psi\frac{\partial P_{s-1}}{\partial y}\right]$$
$$=\left[\frac{\partial P_{s-1}}{\partial \nu}\right]\cos\psi-\frac{\sin\psi}{f\cos\psi}\left[\frac{\partial P_{s-1}}{\partial \psi}\right],$$

so that
$$[P_s]=\left[\frac{\partial P_{s-1}}{\partial \nu}\right]-\frac{\sin\psi}{f\cos^2\psi}\left[\frac{\partial P_{s-1}}{\partial \psi}\right].$$

Now the functions enclosed by square brackets are to be evaluated at $\nu=-x_0$; hence

$$\frac{\partial}{\partial\psi}[P_{s-1}]=\left[\frac{\partial P_{s-1}}{\partial \psi}\right]+f\sin\psi\left[\frac{\partial P_{s-1}}{\partial \nu}\right].$$

Thus the boundary conditions are (4.4.5) and

$$[P_s]=\sec^2\psi\left[\frac{\partial P_{s-1}}{\partial \nu}\right]-\frac{\sin\psi}{f\cos^2\psi}\frac{\partial}{\partial\psi}[P_{s-1}]\quad (s=1,2,\ldots). \tag{4.4.11}$$

From (4.4.5) and (4.4.9) we find at once that

$$P_0=\left(\frac{f\cos\psi}{h}\right)^{\frac{1}{2}}, \tag{4.4.12}$$

which is identical with (4.2.5). Now one can prove by induction that

$$P_s=\sum_{k=0}^{3s}A_{sk}(\psi)\,h^{-k-\frac{1}{2}}. \tag{4.4.13}$$

The coefficients A_{sk} satisfy recurrence relations which are deduced by substituting in (4.4.10) and (4.4.11). We have

$$\left(2\frac{\partial}{\partial\nu}+\frac{2}{h}\right)A_{sk}h^{-k-\frac{1}{2}}=-4kA_{sk}h^{-k-\frac{3}{2}}$$

and
$$\nabla^2(A_{sk}h^{-k-\frac{1}{2}})=[A''_{sk}+(2k+1)^2A_{sk}]\,h^{-k-\frac{5}{2}}$$
$$-[2(k+1)A'_{sk}h_\psi+(k+\tfrac{1}{2})A_{sk}h_{\psi\psi}]\,h^{-k-\frac{7}{2}}$$
$$+\tfrac{1}{4}(2k+1)(2k+5)A_{sk}h_\psi^2h^{-k-\frac{9}{2}},$$

where $h_\psi=f'\cos\psi-3f\sin\psi,\quad h_{\psi\psi}=f''\cos\psi-4f'\sin\psi-3f\cos\psi.$ (4.4.14)

Hence

$$\left.\begin{aligned}-4kA_{sk}=&A''_{s-1,k-1}+(2k-1)^2A_{s-1,k-1}\\&-2(k-1)A'_{s-1,k-2}h_\psi-\tfrac{1}{2}(2k-3)A_{s-1,k-2}h_{\psi\psi}\\&+\tfrac{1}{4}(2k-5)(2k-1)A_{s-1,k-3}h_\psi^2,\end{aligned}\right\} \tag{4.4.15}$$

where $k = 1, 2, \ldots, 3s$ and $A_{s-1,l} = 0$ for $l < 0$. These equations do not give A_{s0}, which must be found by means of (4.4.11). For this purpose it is convenient to put

$$A_{sk} = B_{sk}(f\cos\psi)^{k+\frac{1}{2}}. \tag{4.4.16}$$

Then

$$[P_s] = \sum_{k=0}^{3s} B_{sk}, \tag{4.4.17}$$

and so

$$\sum_{k=0}^{3s} B_{sk} = -\frac{\sec^3\psi}{f}\sum_{k=0}^{3(s-1)}(2k+1)\,B_{s-1,k} - \frac{\sec\psi\tan\psi}{f}\sum_{0}^{3(s-1)} B'_{s-1,k}. \tag{4.4.18}$$

This determines B_{s0} and so also A_{s0} once the B_{sk} with $k \geqslant 1$ have been calculated from (4.4.15) and (4.4.16).

We have

$$A_{00} = (f\cos\psi)^{\frac{1}{2}}. \tag{4.4.19}$$

For $s = 1$ the formulae become

$$\left.\begin{aligned} A_{11} &= -\tfrac{1}{4}(A''_{00} + A_{00}),\\ A_{12} &= \tfrac{1}{4}A'_{00}h_\psi + \tfrac{1}{16}A_{00}h_{\psi\psi},\\ A_{13} &= \tfrac{5}{48}A_{00}h_\psi^2. \end{aligned}\right\} \tag{4.4.20}$$

Instead of using (4.4.18) as it stands we may note that by (4.4.11) and (4.4.12)

$$[P_1] = -\frac{\sec^3\psi}{f}, \tag{4.4.21}$$

so that by (4.4.13)

$$A_{10} = -\frac{A_{11}}{f\cos\psi} - \frac{A_{12}}{f^2\cos^2\psi} - \frac{A_{13}}{f^3\cos^3\psi} - \frac{1}{f^{\frac{1}{2}}\cos^{\frac{5}{2}}\psi}. \tag{4.4.22}$$

The number of terms required for a satisfactory representation of the reflected pulse would have to be estimated in any particular case. The series is of too complicated a form to allow general inferences to be drawn, and numerical evaluation is necessary; one would then be guided by the order of magnitude of the terms. The first coefficient P_1 has a simple interpretation if the incident pulse is a 'unit pressure pulse', that is to say, $p_i = H(ct + x)$. For then cP_1 is the rate of change of the reflected pressure at the reflected front. The equation (4.4.21) shows that at the reflector itself the pressure is initially increasing or decreasing according as the reflector is convex or concave towards the incident front. Thus the pressure tends to increase if the reflected front there is concave, and to decrease if it is convex. If there is a shadow then since $|\psi| \to \frac{1}{2}\pi$ as the shadow boundary is approached along the reflector and f is necessarily positive, the reflected pulse apparently becomes 'infinitely thin' at the shadow boundary. At the same time (4.4.21) indicates that the series solution breaks down at the shadow boundary. It is likely to be unsatisfactory in the neighbourhood of the shadow boundary.

5. Reflexion of a spherical pulse by a paraboloid

A problem to which the method of series expansion has been applied is that of the reflexion of a spherical pressure pulse, emitted at its focus, by a fixed and rigid paraboloid of revolution (Friedlander, 1942*b*). Let the focus be taken as origin and the x-axis along the axis of the paraboloid; its equation is then

$$r^2 = 4a(x+a), \quad r^2 = y^2 + z^2, \tag{4.5.1}$$

where a is the focal length. An alternative form of this equation is

$$R = x + 2a, \quad R = (x^2 + y^2 + z^2)^{\frac{1}{2}}. \tag{4.5.2}$$

Hence an incident front $ct = R$ gives rise to a reflected front $ct = x + 2a$. We take the incident pulse to be

$$p_i = \frac{1}{R} f_0(ct - R), \tag{4.5.3}$$

and expand the reflected pulse as a series,

$$p_r = \sum_{s=0}^{\infty} P_s(x, r)\, f_s(ct - x - 2a). \tag{4.5.4}$$

Then the transport equations are

$$2\frac{\partial P_0}{\partial x} = 0, \quad 2\frac{\partial P_s}{\partial x} = \nabla^2 P_{s-1} = \frac{\partial^2 P_{s-1}}{\partial x^2} + \frac{\partial^2 P_{s-1}}{\partial r^2} + \frac{1}{r}\frac{\partial P_{s-1}}{\partial r} \quad (s \geqslant 1). \tag{4.5.5}$$

The boundary conditions (4.1.6) give

$$\left.\begin{aligned} [P_0] &= \left[\frac{1}{R}\right], \\ [P_1] &= \sec\gamma \left\{\left[\frac{\partial P_0}{\partial n}\right] - \left[\frac{\partial}{\partial n}\frac{1}{R}\right]\right\}, \\ [P_s] &= \sec\gamma \left[\frac{\partial P_{s-1}}{\partial n}\right] \quad (s \geqslant 2), \end{aligned}\right\} \tag{4.5.6}$$

allowing for the fact that the argument of the f_s is $ct-x-2a$ rather than $t-(x+2a)/c$. The angle γ between the reversed incident ray and the normal drawn into physical space is here equal to the angle between this normal and the x-axis, so that

$$\tan\gamma = -\mathrm{d}x/\mathrm{d}r = -r/2a.$$

Thus
$$\sec\gamma\left[\frac{\partial}{\partial n}\right]=\left[\frac{\partial}{\partial x}-\frac{r}{2a}\frac{\partial}{\partial r}\right].$$

Hence
$$\sec\gamma\left[\frac{\partial}{\partial n}\frac{1}{R}\right]=\left[\frac{r^2-2ax}{2aR^3}\right]=\left[\frac{2a(x+2a)}{2aR^3}\right]=\left[\frac{1}{R^2}\right]=\left(\frac{4a}{4a^2+R^2}\right)^2,$$

and so the equations (4.5.6) are in terms of x and r

$$\left.\begin{aligned}
[P_0] &= \frac{4a}{4a^2+r^2},\\
[P_1] &= \left(\frac{4a}{4a^2+r^2}\right)^2+\left[\frac{\partial P_0}{\partial x}-\frac{r}{2a}\frac{\partial P_0}{\partial r}\right],\\
[P_s] &= \left[\frac{\partial P_{s-1}}{\partial x}-\frac{r}{2a}\frac{\partial P_{s-1}}{\partial r}\right] \quad (s\geqslant 2).
\end{aligned}\right\} \tag{4.5.7}$$

Let us introduce two new independent variables,

$$u=1+\frac{r^2}{4a^2},\quad v=\frac{x}{a}+1-\frac{r^2}{4a^2}. \tag{4.5.8}$$

These are both non-dimensional. If F is any function of x and r, then

$$\frac{\partial F}{\partial x}=\frac{1}{a}\frac{\partial F}{\partial v},\quad \frac{\partial F}{\partial r}=\frac{r}{2a^2}\left(\frac{\partial F}{\partial u}-\frac{\partial F}{\partial v}\right),$$

$$\frac{\partial^2 F}{\partial x^2}=\frac{1}{a^2}\frac{\partial^2 F}{\partial r^2},\quad \frac{\partial^2 F}{\partial r^2}=\frac{1}{2a^2}\left(\frac{\partial F}{\partial u}-\frac{\partial F}{\partial v}\right)+\frac{r^2}{4a^4}\left(\frac{\partial^2 F}{\partial u^2}-2\frac{\partial^2 F}{\partial u\,\partial v}+\frac{\partial^2 F}{\partial v^2}\right),$$

Hence the transport equations (4.5.5) become

$$\left.\begin{aligned}
&\frac{\partial P_0}{\partial v}=0,\\
&2a\frac{\partial P_s}{\partial v}=(u-1)\left(\frac{\partial^2 P_{s-1}}{\partial u^2}-2\frac{\partial^2 P_{s-1}}{\partial u\,\partial v}\right)\\
&\qquad +u\frac{\partial^2 P_{s-1}}{\partial v^2}+\frac{\partial P_{s-1}}{\partial u}-\frac{\partial P_{s-1}}{\partial v} \quad (s\geqslant 1),
\end{aligned}\right\} \tag{4.5.9}$$

and the boundary conditions (4.5.7),

$$\left.\begin{aligned} a[P_0]_{v=0} &= \frac{1}{u}, \\ a^2[P_1]_{v=0} &= \frac{1}{u^2} + a\left[u\frac{\partial P_0}{\partial v} - (u-1)\frac{\partial P_0}{\partial u}\right]_{v=0}, \\ a[P_s]_{v=0} &= \left[u\frac{\partial P_{s-1}}{\partial v} - (u-1)\frac{\partial P_{s-1}}{\partial u}\right]_{v=0} \quad (s \geqslant 2). \end{aligned}\right\} \tag{4.5.10}$$

Since P_0 is independent of v we have immediately that

$$aP_0 = \frac{1}{u}. \tag{4.5.11}$$

Hence
$$a^2[P_1]_{v=0} = \frac{1}{u^2} + \frac{u-1}{u^2} = \frac{1}{u}$$

and
$$2a^2\frac{\partial P_1}{\partial v} = \frac{2(u-1)}{u^3} - \frac{1}{u^2} = \frac{1}{u^2} - \frac{2}{u^3}.$$

Thus
$$a^2P_1 = \frac{1}{u} + v\left(\frac{1}{2u^2} - \frac{1}{u^3}\right). \tag{4.5.12}$$

It is obvious that $a^{s+1}P_s$ is a polynomial in v of degree s whose coefficients are polynomials in $1/u$. It can be shown by induction that

$$a^{s+1}P_s = \sum_{k=1}^{s} a_{s0k}u^{-k} + \sum_{l=1}^{s}\sum_{k=1}^{s+1} a_{slk}v^l u^{-l-k}, \tag{4.5.13}$$

where the a_{slk} are constants. These can be computed from recurrence relations that are obtained by substituting (4.5.13) in (4.5.9) and (4.5.10); they are

$$a_{s0k} = a_{s-1,0,k} + ka_{s-1,0,k} - (k-1)\,a_{s-1,0,k-1}, \tag{4.5.14}$$

$$a_{slk} = -\frac{(l+k-1)(l+k-2)}{2l}a_{s-l,l-1,k-1} + \frac{(l+k-1)^2}{2l}a_{s-1,l-1,k}$$
$$-(l+k-1)\,a_{s-1,l,k-1} + (l+k-\tfrac{1}{2})\,a_{slk} + \tfrac{1}{2}(l+1)\,a_{s,l+1,k}. \tag{4.5.15}$$

These relations are valid for $s \geqslant 2$ and $0 \leqslant l \leqslant s$, $1 \leqslant k \leqslant s+1$, provided that such terms as $a_{s-1,0,s}$, $a_{s-1,s,k}$ and $a_{s-1,l,s+1}$ on the right-hand side are replaced by zero. A table of these coefficients for $s = 2, 3, \ldots, 6$ is given in the paper already referred to.

Once these coefficients have been determined, the values of $P_0, \ldots, P_6$ can be calculated. The calculations which were made indicated that the series converges more rapidly at points on the paraboloid than on the axis, and that its rate of convergence decreases as x increases. The reflected pressure pulse due to the incident pulse $H(ct-R)/R$ has one interesting feature which can already be

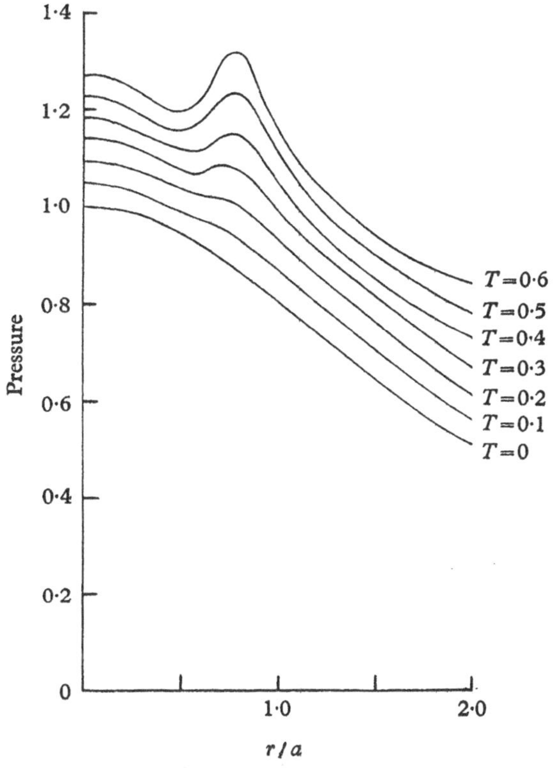

Fig. 4.6. Distribution of the reflected pressure pulse across a paraboloid of revolution.

inferred from the behaviour of P_0 and P_1. By (4.5.11), the initial pressure rise is greatest on the axis and falls off as r increases. On the other hand, P_1, which in this case is $\partial p_r/c\,\partial t$ evaluated at the reflected front, is positive on the paraboloid and negative on the axis for $x > a$. This suggests that a secondary pressure maximum will be formed on a ring $r = \text{const}$. The calculations bear this out; an example is shown in fig. 4.6 where ap_r is plotted against r/a for various values of $T=(ct-x-2a)/a$ in the focal plane ($x=0$). This conclusion is borne out qualitatively by some measurements made by Anderson (1952).

6. Reflexion of a plane pulse by a convex paraboloid

Another problem involving parabolic boundaries is the reflexion of a plane pulse by a convex paraboloid of revolution, or by a parabolic cylinder. The solution in the harmonic case is due to Lamb (1906); the pulse case has been discussed by Friedlander (1941) and by Chester (1952). We shall here consider the reflexion by a paraboloid in detail; the corresponding results for the parabolic cylinder will be stated briefly at the end of § 7.

Let the equation of the paraboloid be now

$$r^2=4a(a-x), \tag{4.6.1}$$

where a is again the focal length. Then also $R=2a-x$ on the paraboloid. Hence the incident fronts $ct+x=0$ are reflected as the concentric spheres $R=ct+2a$ ($ct>a$). It is more convenient here to work with the velocity potential Φ than with the pressure. Let

$$\Phi_i=f(ct+x) \tag{4.6.2}$$

be the incident plane pulse, where $f(ct)=0$ for $t\leqslant 0$. The velocity potential Φ_r of the reflected pulse satisfies the wave equation, the boundary condition

$$\left[\frac{\partial\Phi_r}{\partial n}\right]=-\left[\frac{\partial\Phi_i}{\partial n}\right] \tag{4.6.3}$$

on the paraboloid, and vanishes for $ct\leqslant R-2a$. Now it can be shown that Φ_r depends only on the time delays $ct+x$, $ct-R+2a$. This is equivalent to saying that Φ_r is a function of $ct-R+2a$ and $R+x$ only. Hence Φ_r takes the same succession of values on every paraboloid confocal with the reflector. This is the characteristic feature of the problem.

To prove the assertion, let us assume that

$$\Phi_r=F(\xi,\eta), \tag{4.6.4}$$

where

$$\xi=ct+x,\quad \eta=ct-R+2a, \tag{4.6.5}$$

so that $\xi=\eta$ on the paraboloid. Then

$$\frac{\partial\Phi_r}{\partial x}=\frac{\partial F}{\partial\xi}-\frac{x}{R}\frac{\partial F}{\partial\eta},\quad \frac{\partial\Phi_r}{\partial r}=-\frac{r}{R}\frac{\partial F}{\partial\eta},\quad \frac{1}{c}\frac{\partial\Phi}{\partial t}=\frac{\partial F}{\partial\xi}+\frac{\partial F}{\partial\eta},$$

and
$$\nabla^2\Phi_r = \frac{\partial^2 F}{\partial \xi^2} - 2\frac{x}{R}\frac{\partial^2 F}{\partial \xi \partial \eta} + \frac{\partial^2 F}{\partial \eta^2} - \frac{2}{R}\frac{\partial F}{\partial \eta},$$
$$\frac{1}{c^2}\frac{\partial^2 \Phi_r}{\partial t^2} = \frac{\partial^2 F}{\partial \xi^2} + 2\frac{\partial^2 F}{\partial \xi \partial \eta} + \frac{\partial^2 F}{\partial \eta^2}.$$

Since $R+x=\xi-\eta+2a$, it follows that Φ_r will be a solution of the wave equation if F satisfies
$$\frac{\partial^2 F}{\partial \xi \partial \eta} + (\xi-\eta+2a)\frac{\partial F}{\partial \eta} = 0. \tag{4.6.6}$$

But this equation can be integrated. We have first that
$$\frac{\partial F}{\partial \eta} = \frac{g(\eta)}{\xi-\eta+2a},$$
where g is arbitrary, and so can take
$$F(\xi,\eta) = \int_0^\eta \frac{g(\zeta)}{\xi-\zeta+2a}\,\mathrm{d}\zeta. \tag{4.6.7}$$

If we assume that $g(\zeta)=0$ for $\zeta \leqslant 0$ then $F=0$ for $\eta \leqslant 0$, that is to say, for $ct \leqslant R-2a$; this is one of the conditions to be satisfied. It remains to determine g by substituting (4.6.7) in the boundary condition (4.6.3). Now this condition is equivalent to
$$\left[\frac{\partial \Phi_r}{\partial x} + \frac{r}{2a}\frac{\partial \Phi_r}{\partial r}\right] = -\left[\frac{\partial \Phi_i}{\partial x} + \frac{r}{2a}\frac{\partial \Phi_i}{\partial r}\right] = -f'(ct+x)$$
and
$$\left[\frac{\partial \Phi_r}{\partial x} + \frac{r}{2a}\frac{\partial \Phi_r}{\partial r}\right] = \left[\frac{\partial F}{\partial \xi} - \frac{x}{R}\frac{\partial F}{\partial \eta} - \frac{r^2}{2aR}\frac{\partial F}{\partial \eta}\right]_{\xi=\eta} = \left[\frac{\partial F}{\partial \xi} - \frac{\partial F}{\partial \eta}\right]_{\xi=\eta}.$$

Hence
$$\left[\frac{\partial F}{\partial \eta} - \frac{\partial F}{\partial \xi}\right]_{\xi=\eta} = f'(\eta),$$
and if we substitute (4.6.7),
$$\frac{g(\eta)}{2a} + \int_0^\eta \frac{g(\zeta)}{(\eta+2a-\zeta)^2}\,\mathrm{d}\zeta = f'(\eta). \tag{4.6.8}$$

This can be written as an integral equation for the total (excess) pressure on the paraboloid, p^*. This is the sum of the incident pressure pulse
$$p_i = \left[\frac{\partial \Phi_i}{\partial t}\right] = c[f'(ct+x)] = cf'(\eta), \tag{4.6.9}$$

and of the reflected pressure pulse

$$p_r = \left[\frac{\partial \Phi}{\partial t}\right] = c\left[\frac{\partial F}{\partial \xi} + \frac{\partial F}{\partial \eta}\right]_{\xi=\eta} = \frac{c}{2a} g(\eta) - c\int_0^\eta \frac{g(\zeta)}{(\eta+2a-\zeta)^2}\,d\zeta.$$

Hence by (4.6.8)

$$p^* = p_i + p_r = \frac{c}{a} g(\eta). \tag{4.6.10}$$

Denoting the incident pressure pulse $cf'(ct+x)$ by $p_0(ct+x)$, the integral equation (4.6.8) is therefore

$$p^*(\eta) + 2a\int_0^\eta \frac{p^*(\zeta)}{(\eta+2a-\zeta)^2}\,d\zeta = 2p_0(\eta). \tag{4.6.11}$$

The reflected pressure pulse at points not on the paraboloid can also be written in terms of p^*,

$$\begin{aligned} p_r = \frac{\partial \phi}{\partial t} &= \frac{a}{\xi-\eta+2a} p^*(\eta) - a\int_0^\eta \frac{p^*(\zeta)}{(\xi-\zeta+2a)^2}\,d\zeta \\ &= \frac{a}{R+x} p^*(ct-R+2a) - a\int_0^{ct-R+2a} \frac{p^*(\zeta)}{(ct+x+2a-\zeta)^2}\,d\zeta. \end{aligned} \tag{4.6.12}$$

The integral equation (4.6.11) is a Volterra equation of the second kind. Its solution may be written as

$$p^*(\eta) = 2p_0(\eta) - 2\int_0^\eta k(\eta-\zeta)\,p_0(\zeta)\,d\zeta, \tag{4.6.13}$$

where the 'resolving kernel' k satisfies

$$k(\eta) + 2a\int_0^\eta \frac{k(\zeta)}{(\eta-\zeta+2a)^2}\,d\zeta = \frac{2a}{(\eta+2a)^2}. \tag{4.6.14}$$

For an incident unit pressure pulse $H(ct+x)$ the pressure $P(\eta)$ on the paraboloid is by (4.6.13)

$$P(\eta) = 2 - 2\int_0^\eta k(\zeta)\,d\zeta. \tag{4.6.15}$$

Hence in the general case

$$p^*(\eta) = 2p_0(\eta) - \int_0^\eta P'(\eta-\zeta)\,p_0(\zeta)\,d\zeta = \int_0^\eta p_0(\eta-\zeta)\,dP(\zeta). \tag{4.6.16}$$

This is an example of the superposition principle of § 3·8. $P(\eta)$ satisfies the integral equation

$$P(\eta) + 2a\int_0^\eta \frac{P(\zeta)}{(2a+\eta-\zeta)^2}\,d\zeta = 2H(\eta) \tag{4.6.17}$$

and depends only on η/a. One can prove that P decreases steadily from $P(0)=2$ to unity as $\eta\to\infty$, so that in time the pressure on the paraboloid reduces to the incident pressure.† In the first place, (4.6.17) implies that $0<P(\eta)\leqslant 2$ for $\eta\geqslant 0$. For P is obviously positive at first, and (4.6.17) implies that $P\leqslant 2$ when $P\geqslant 0$. Suppose now that $P(\eta)>0$ in $0\leqslant\eta<Y$; then it can be shown at once that $P(Y)=0$ implies a contradiction. For

$$0=P(Y)=2-2a\int_0^Y\frac{P(\zeta)}{(Y-\zeta+2a)^2}d\zeta\geqslant 2-4a\int_0^Y\frac{d\zeta}{(Y-\zeta+2a)^2}$$
$$=\frac{4a}{4a+Y}>0,$$

which is impossible; hence $P(Y)>0$ for all $Y>0$.

Secondly, we can prove that if P tends to a limit as $\eta\to\infty$ then this limit is unity. For suppose that $P(\eta)\to A$ as $\eta\to\infty$. Then given any $\epsilon>0$ there is a $Y>0$ such that $|P(\eta)-A|<\epsilon$ for $\eta>Y$. Hence for $\eta>Y$

$$\left|2a\int_0^\eta\frac{P(\zeta)}{(\eta-\zeta+2a)^2}d\zeta-A\right|\leqslant\left|2a\int_0^Y\frac{P(\zeta)}{(\eta-\zeta+2a)^2}d\zeta\right|$$
$$+\left|2a\int_Y^\eta\frac{P(\zeta)-A}{(\eta-\zeta+2a)^2}d\zeta\right|$$
$$+|A|\left|2a\int_Y^\eta\frac{d\zeta}{(\eta-\zeta+2a)^2}-1\right|$$
$$\leqslant 4a\left(\frac{1}{\eta+2a-Y}-\frac{1}{2a+\eta}\right)$$
$$+2a\epsilon\left(\frac{1}{2a}-\frac{1}{2a+\eta-Y}\right)+\frac{2a|A|}{2a+\eta-Y}$$
$$<2\epsilon$$

for η sufficiently large. Thus the assumption that $\lim_{\eta\to\infty}P(\eta)$ exists implies that

$$\lim_{\eta\to\infty}\left\{\int_0^\eta\frac{2aP(\zeta)}{(\eta-\zeta+2a)^2}d\zeta\right\}=\lim_{\eta\to\infty}P(\eta),$$

and so by (4.6.17) $\lim_{\eta\to\infty}P(\eta)=1$.

† For $ct>-x$ the incident field is a uniform flow in the negative x-direction. It can be shown that the total velocity field is asymptotic to the steady irrotational flow past the paraboloid. The pressure then differs from unity by an amount proportional to the square of the particle velocity, and this is neglected in the acoustic approximation.

It remains to establish that P is a decreasing function of η (this will of course ensure the existence of $\lim_{\eta\to\infty} P(\eta)$). We shall do this by obtaining an integral representation of P which is of intrinsic interest. Since P is bounded, its Laplace transform

$$\bar{P}(s) = \int_0^\infty e^{-s\eta} P(\eta)\, d\eta \tag{4.6.18}$$

exists. We can therefore take the Laplace transform of (4.6.17) which is

$$\bar{P}(s)\{1+\bar{K}(s)\} = \frac{2}{s},$$

where $$\bar{K}(s) = 2a\int_0^\infty \frac{e^{-\eta s}}{(2a+\eta)^2}\, d\eta = 1 + 2as\, e^{2s}\, \mathrm{Ei}(-2as),$$

Ei being the exponential integral,

$$\mathrm{Ei}(-z) = -\int_z^\infty \frac{e^{-\zeta}}{\zeta}\, d\zeta. \tag{4.6.19}$$

Hence in the first place

$$P(\eta) = \frac{1}{2\pi i}\int_{C-i\infty}^{C+i\infty} \frac{e^{s\eta}}{1+as\, e^{2as}\, \mathrm{Ei}(-2as)} \frac{ds}{s}, \tag{4.6.20}$$

with $C>0$. Now (4.6.19) defines $\mathrm{Ei}(-2as)$ in the complex s-plane cut along the negative real axis. For large s,

$$1 + as\, e^{2as}\, \mathrm{Ei}(-2as) \sim \tfrac{1}{2},$$

and one can show that $1 + as\, e^{2as}\, \mathrm{Ei}(-2as)$ has no zeros in the cut s-plane. Hence the contour can be deformed into the loop $(-\infty, 0+)$ One can write (4.6.19) as

$$\mathrm{Ei}(-z) = -\int_z^1 \frac{e^{-\zeta}-1}{\zeta}\, d\zeta - \int_1^\infty \frac{e^{-\zeta}}{\zeta}\, d\zeta + \log z.$$

Also $$\log\gamma = \int_0^1 \frac{1-e^{-\zeta}}{\zeta}\, d\zeta - \int_1^\infty \frac{e^{-\zeta}}{\zeta}\, d\zeta,$$

where γ is Euler's constant. Hence

$$\mathrm{Ei}(-z) = -\int_0^z \frac{1-e^{-\zeta}}{\zeta}\, d\zeta + \log\gamma z. \tag{4.6.21}$$

Contracting the loop $(-\infty, 0+)$ to the upper and lower sides of the cut $\arg s = \pi$, $\arg s = -\pi$ respectively, with a small circle about the origin, we obtain therefore

$$P(\eta) = 1 + \int_0^\infty \frac{e^{-\sigma(\eta+2a)}}{[1 - a\sigma\, e^{-2a\sigma}\,\overline{\mathrm{Ei}}(2a\sigma)]^2 + \pi^2 a^2 \sigma^2\, e^{-4a\sigma}}\, d\sigma, \quad (4.6.22)$$

where

$$\overline{\mathrm{Ei}}(z) = \int_0^z \frac{e^\zeta - 1}{\zeta}\, d\zeta + \log \gamma z. \quad (4.6.23)$$

It is obviously legitimate to differentiate under the integral sign so that

$$k(\eta) = -\tfrac{1}{2}P'(\eta) = \frac{1}{2}\int_0^\infty \frac{\sigma\, e^{-\sigma(\eta+2a)}}{[1 - a\sigma\, e^{-2a\sigma}\,\overline{\mathrm{Ei}}(2a\sigma)]^2 + \pi^2 a^2 \sigma^2\, e^{-4a\sigma}}\, d\sigma. \quad (4.6.24)$$

Hence $k > 0$, $P' < 0$, P is strictly decreasing and, as we have shown this implies that $\lim_{\eta\to\infty} P(\eta) = 1$.

Although $\overline{\mathrm{Ei}}$ has been tabulated, (4.6.22) and (4.6.24) are not particularly convenient for the computation of P or k unless η is large, and it is better to compute these functions by other means. One method which was used in Friedlander (1941) employs the series solutions of (4.6.14) or (4.6.17) obtained by iteration, which converge for all η. It is rather laborious and other methods are preferable in practice.† The graph of $P(\eta)$ is shown in fig. 4.7 for $0 \leqslant \eta \leqslant 4a$. The reflected pressure pulse $P(\eta) - 1$ is reduced by about 50% in the time a/c taken by a wave front to travel a distance equal to the focal length of the reflector. The reflected pressure pulse is $P(\eta) - 1$ at the paraboloid, while at a large distance from the paraboloid it is

$$\frac{a}{R+x} P(\eta) + O\left(\frac{1}{R^2}\right)$$

by (4.6.12). This illustrates the general rule that a pulse is distorted as it propagates; in this example the reflected pulse dies down more slowly at a large distance from the reflector than it does at the reflector itself.

† See L. Fox and E. T. Goodwin (1953). These authors use (4.6.14), with $a = 1$, as an illustrative example of a Volterra equation (loc. cit. pp. 523–30). The table of $\frac{1}{2}k(\zeta)$ (in the notation of the text) given in Friedlander (1941) is shown to be inaccurate, the error in the fourth decimal at $\eta = 1, 2, 3$ and 4 being 4, 5, 13 and 25 units respectively. The graph of P shown in fig. 4.7 has nevertheless been computed from this table, since the scale of the figure only allows two decimals to be taken into account.

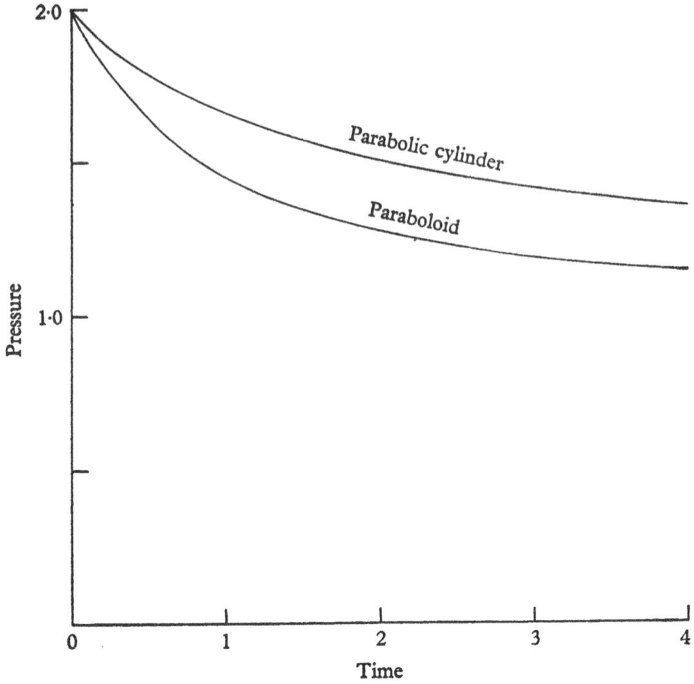

Fig. 4.7. Total excess pressure on a convex paraboloid of revolution and on a parabolic cylinder, due to an incident unit pressure pulse.

7. Series expansion of the reflected pulse

We have so far not used the series expansion method in this example at all. This gives an alternative approach, but the most interesting feature is perhaps that the radius of convergence of the series can be determined.

The series expansion of the reflected pulse in terms of the repeated integrals of p_0 can be obtained as follows. Let

$$p^*(\eta) = \sum_{s=0}^{\infty} A_s p_s(\eta), \tag{4.7.1}$$

where the A_s are constants and

$$p_s(\eta) = \int_0^\eta p_{s-1}(\zeta)\,d\zeta \quad (s = 1, 2, \ldots). \tag{4.7.2}$$

Then
$$\frac{1}{m!}\int_0^\eta (\eta-\zeta)^m p_s(\zeta)\,d\zeta = p_{s+m+1}(\eta). \tag{4.7.3}$$

If we substitute (4.7.1) in (4.6.11) and expand the kernel of the integral equation as a power series in $\eta-\zeta$,

$$\frac{2a}{(\eta+2a-\zeta)^2}=\sum_{m=0}^{\infty}(-1)^m\frac{(m+1)!}{m!}\frac{(\eta-\zeta)^m}{(2a)^{m+1}}, \tag{4.7.4}$$

we obtain therefore

$$\sum_{s=0}^{\infty}A_s p_s(\eta)+\sum_{s=0}^{\infty}\sum_{m=0}^{\infty}(-1)^m\frac{(m+1)!}{(2a)^{m+1}}A_s p_{s+m+1}(\eta)-2p_0(\eta)=0.$$

This will hold only if the coefficients of $p_0, p_1, \ldots$ vanish,

$$A_0=2,\quad A_s+\sum_{m=1}^{s}(-1)^{m-1}\frac{m!}{(2a)^m}A_{s-m}=0\quad (s=1,2,\ldots).$$

If we put

$$B_s=(-2a)^s A_s, \tag{4.7.5}$$

then these recurrence equations become

$$B_0=2,\quad B_s=\sum_{m=1}^{s}m!\,B_{s-m}. \tag{4.7.6}$$

The series (4.7.1) with the coefficients defined by (4.7.5) and (4.7.6) converges in general for $0\leqslant\eta<2a$. To prove this we observe that by (4.7.6)

$$sB_{s-1}-B_s=\sum_{m=0}^{s-2}[s(s-m-1)!-(s-m)!]\,B_m-B_{s-1}.$$

The B_s are obviously all positive; hence

$$(s+1)\,B_{s-1}>B_s,$$

and since $B_0=2$ it follows that

$$B_s\leqslant 2(s+1)!. \tag{4.7.7}$$

Also by (3.6.9)

$$|p_s(\eta)|\leqslant\frac{Mk!}{(s+k)!}\eta^{s+k},$$

provided that $|p_0|\leqslant M\eta^k$, where M and k are constants, $k>-1$. Hence (4.7.1) is dominated by the series

$$2Mk!\,\eta^k\sum_{s=0}^{\infty}\frac{(s+1)!}{(s+k)!}\left(\frac{\eta}{2a}\right)^s,$$

which converges for $0\leqslant\eta<2a$. Since (4.7.4) also converges for $0\leqslant\zeta\leqslant\eta<2a$ and both series converge uniformly in any sub-interval of $(0, 2a)$, the method which we have used formally is justified and (4.7.1) is a valid expansion of the solution of the integral equation for $0 \leqslant \eta < 2a$.

It remains to calculate p_r. This is given by (4.6.12) which we can write as

$$p_r = \frac{a}{R+x} p^*(\eta) - a \int_0^\eta p^*(\zeta) \left\{ \sum_{m=0}^{\infty} (-1)^m \frac{(m+1)!}{m!} \frac{(\eta-\zeta)^m}{(R+x)^{m+2}} \right\} d\zeta$$

for $0 \leqslant \eta < 2a$, since $R+x \geqslant 2a$. Using (4.7.1), (4.7.5) and (4.7.3) we therefore find

$$p_r = \frac{a}{R+x} \sum_{s=0}^{\infty} \frac{B_s}{(-2a)^s} p_s(\eta)$$
$$- \sum_{s=0}^{\infty} \sum_{m=0}^{\infty} (-1)^m \frac{(m+1)!\, a}{(R+x)^{m+2}} \frac{B_s}{(-2a)^s} p_{s+m+1}(\eta),$$

that is to say,

$$p_r = \frac{1}{2} \sum_{s=0}^{\infty} \left(\frac{-1}{2a} \right)^s p_r(\eta) \sum_{m=0}^{s} m!\, B_{s-m} \left(\frac{2a}{R+x} \right)^{m+1}. \qquad (4.7.8)$$

Since the B_s are positive and $R+x \geqslant 2a$ in physical space,

$$\sum_0^s m!\, B_{s-m} \left(\frac{2a}{R+x} \right)^{m+1} \leqslant \Sigma\, m!\, B_{n-m} = 2B_n,$$

so that this series also converges for $0 \leqslant \eta < 2a$ in physical space.

The fact that (4.7.1) and (4.7.8) have a finite range of convergence is from a physical point of view purely accidental; it arises from a singularity of the solution when extended into the 'non-physical' part of space. The applicability of the series expansions considered in this chapter is liable to be limited in this way in many cases.

The case where the reflector is the parabolic cylinder

$$y^2 = 4a(a-x) \qquad (4.7.9)$$

can be treated in the same way (Friedlander, 1941). One has only to replace the distance R from the focus of the paraboloid by the distance $r = (x^2+y^2)^{\frac{1}{2}}$ from the focal line of the parabolic cylinder. The integral equation for the total excess pressure p^* on the reflector is found to be

$$p^*(\eta) + \left(\frac{a}{2} \right)^{\frac{1}{2}} \int_0^\eta \frac{p^*(\zeta)}{(\eta - \zeta + 2a)^{\frac{3}{2}}} d\zeta = 2p_0(\eta), \qquad (4.7.10)$$

and the reflected pressure pulse is

$$p_r = \left\{\frac{a}{2(r+x)}\right\}^{\frac{1}{2}} p^*(\eta) - \frac{1}{2}\left(\frac{a}{2}\right)^{\frac{1}{2}} \int_0^{\eta} \frac{p^*(\zeta)}{(r+x+\eta-\zeta)^{\frac{3}{2}}} d\zeta. \quad (4.7.11)$$

For $0 \leqslant \eta < 2a$, the integral equation (4.6.10) is satisfied by the series

$$p(\eta) = \sum_{s=0}^{\infty} \left(\frac{-1}{4a}\right)^s C_s p_s(\eta), \quad (4.7.12)$$

where the coefficients C_s must be calculated by means of the recurrence equations

$$C_0 = 2, \quad C_s = \sum_{m=1}^{s} 1.3\ldots\ldots(2m-1)\, C_{s-m}. \quad (4.7.13)$$

The series expansion of the reflected pressure pulse is

$$p_r = \frac{1}{2} \sum_{s=0}^{\infty} \left(\frac{-1}{4a}\right)^s p_s(\eta) \sum_{m=0}^{s} 1.3\ldots\ldots(2m-1)\, C_{s-m} \left(\frac{2a}{r+x}\right)^{m+\frac{1}{2}}. \quad (4.7.14)$$

The function P which results when the incident pressure pulse is $H(ct+x)$ is also shown in fig. 4.7.

8. The refraction of a spherical pulse at a plane interface

Geometrical acoustics has been applied by Friedrichs and Keller (1955) to the refraction of a spherical pulse at a plane interface. The solution of this problem can be obtained as the inverse Laplace transform or Fourier transform of a contour integral (Garnir, 1953; Gerjouy, 1953) or by an ingenious method due to John (1954). But it is difficult to disentangle the characteristic features of the disturbance from these solutions, while the method of Friedrichs and Keller is simple. It does, however, only give a partial solution. The strengths of the reflected and transmitted acoustic shocks are determined, also—and this is the particular merit of the method—the frontal behaviour of the diffracted pulse which is formed when the velocity of sound in the medium containing the source is less than the velocity of sound in the other medium.

Let the half-spaces $x > 0$ and $x < 0$ contain different homogeneous media; we denote the density and velocity of sound by ρ_1 and c_1 respectively in $x > 0$, and by ρ_2 and c_2 respectively in $x < 0$.

The source is to be situated in the 'upper' medium 1 at a point S on the x-axis at distance h from the interface. The incident pressure pulse is $H(ct-R)/R$, where R is the distance from the source $\{(x-h)^2+r^2\}^{\frac{1}{2}}$; r is the distance from the axis. Because of the axial symmetry we can work in a plane through the x-axis. An incident ray SA making an angle α with the negative x-axis meets the interface at the point A: $x=0$, $r=h\tan\alpha$. At A there is a reflected ray AB at an angle α to the positive x-axis and a refracted (transmitted) ray AC at an angle β to the negative x-axis (fig. 4.8). According to the refraction law of geometrical optics,

$$\frac{\sin\beta}{\sin\alpha}=\frac{c_2}{c_1}. \tag{4.8.1}$$

If $c_2 \leqslant c_1$, β is real for all values of α, that is to say, for $0 \leqslant \alpha \leqslant \frac{1}{2}\pi$. But when $c_2 > c_1$ there is a *critical angle* ϵ,

$$\sin\epsilon=\frac{c_1}{c_2}, \tag{4.8.2}$$

and (4.8.1) gives $\sin\beta > 1$ for $\alpha > \epsilon$ so that then no transmitted ray exists.

Let us first assume that $\sin\beta < 1$. Then the strengths $P^{(r)}$, $P^{(t)}$ of the reflected and transmitted acoustic shocks can be determined at once. The pressure jump must be the same on either side of the interface,

$$\frac{1}{R_0}+P^{(r)}(A)=P^{(t)}(A), \quad R_0=h\sec\alpha.$$

Also, the discontinuous variation of the component of particle velocity normal to the interface must be the same on either side; hence by (3.3.9)

$$\frac{1}{\rho_1 c_1}\left\{\frac{\cos\alpha}{R_0}-P^{(r)}(A)\cos\alpha\right\}=\frac{P^{(t)}(A)\cos\beta}{\rho_2 c_2}.$$

From these two equations we deduce at once that

$$\left.\begin{aligned} P^{(r)}(A)&=\frac{1}{R_0}\frac{\rho_2c_2\cos\alpha-\rho_1c_1\cos\beta}{\rho_2c_2\cos\alpha+\rho_1c_1\cos\beta},\\ P^{(t)}(A)&=\frac{1}{R_0}\frac{2\rho_2c_2\cos\alpha}{\rho_2c_2\cos\alpha+\rho_1c_1\cos\beta}.\end{aligned}\right\} \tag{4.8.3}$$

These formulae are similar to those which determine the corresponding acoustic shock strengths for an incident plane pulse travelling in a direction that makes an angle α with the negative x-axis.

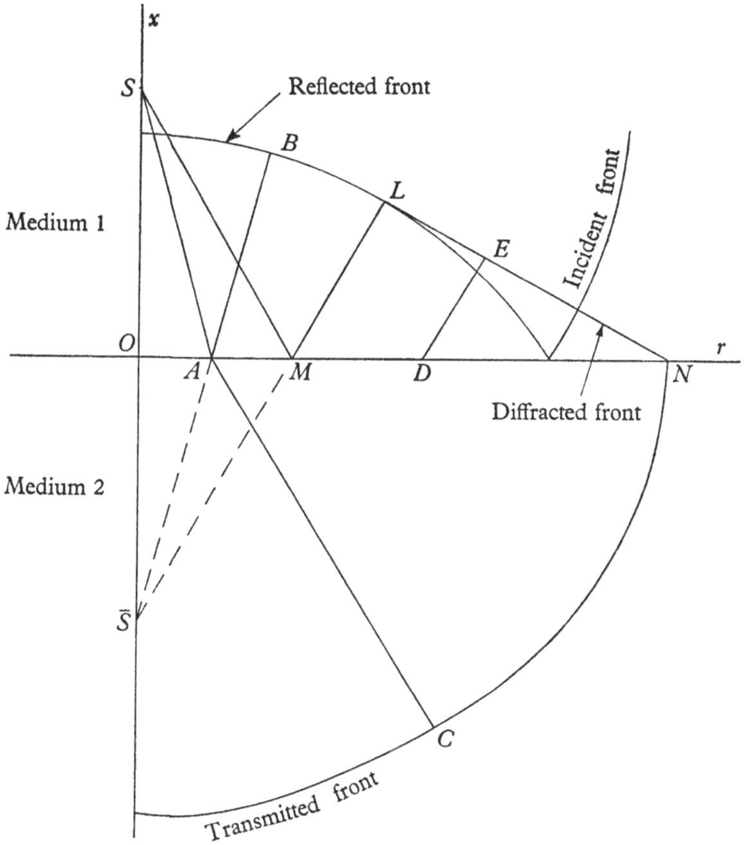

Fig. 4.8. Wave fronts arising in the refraction of a spherical front at a plane interface.

The strength of the reflected shock at the point B is then

$$P^{(r)}(B) = \frac{1}{\bar{R}} P^{(r)}(A), \tag{4.8.4}$$

where $\bar{R}$ is the distance of B from the image $\bar{S}$ of the source in the interface. If the coordinates of B are (x, r) then $\bar{R} = \{(x+h)^2 + r^2\}^{\frac{1}{2}}$, $\tan\alpha = r/(h+x)$, and the distance of A from the axis is $hr/(h+x)$;

hence (4.7.4) gives the reflected shock strength explicitly as a function of x and r. But it should be noted that if $c_2 > c_1$ then it is only valid for $\alpha < \epsilon$.

To calculate the strength of the transmitted shock we can use (4.1.8) again. The principal radii of curvature of the transmitted front are its radius of curvature F in a meridional plane and the distance from the axis measured along the refracted ray. Hence

$$P^{(t)}(C) = P^{(t)}(A)\left\{\frac{F}{F+\sigma}\frac{h\tan\alpha}{h\tan\alpha+\sigma\sin\beta}\right\}^{\frac{1}{2}}, \qquad (4.8.5)$$

where σ is the distance AC. Now adjacent rays at angles α, $\alpha+\delta\alpha$ are transmitted as rays which make an angle $\delta\beta$ with each other; they intercept a segment $\delta(h\tan\alpha)$ on the interface whose projection normal to the transmitted ray is $\delta(h\tan\alpha)\cos\beta$. Thus

$$F\delta\beta = \delta(h\tan\alpha)\cos\beta,$$

and so by (4.7.1) and (4.7.2)

$$F = \frac{c_1}{c_2}h\sec^3\alpha\cos^2\beta = h\sin\epsilon\sec^3\alpha\cos^2\beta, \qquad (4.8.6)$$

where the second form only applies when $c_2 > c_1$.

When $c_2 \leqslant c_1$, the formulae (4.8.3)–(4.8.6) constitute the geometrical acoustics approximation to the solution of our problem. But when $c_2 > c_1$ the refracted front runs ahead of the incident front on the interface (this is the case illustrated in fig. 4.8). Its distance from the axis at t is obtained by going along the critical ray SMN:

$$t = \frac{h\sec\epsilon}{c_1} + \frac{r-h\tan\epsilon}{c_2}. \qquad (4.8.7)$$

The incident front meets the interface along the circle

$$c_1 t = (r^2+h^2)^{\frac{1}{2}},$$

and for given t the value of r calculated from this is less than that obtained from (4.7.7) if $t > h/c_1\cos\epsilon$. A wave front must therefore propagate back into the first medium which joins (4.6.7) on $x=0$. This is the conical front

$$c_1 t = r\sin\epsilon + (h+x)\cos\epsilon, \qquad (4.8.8)$$

since by (4.8.1) and (4.8.2)

$$\frac{\sin\epsilon}{c_1}=\frac{1}{c_2},\quad \frac{h\cos\epsilon}{c_1}=\frac{h\sec\epsilon}{c_1}-\frac{h\tan\epsilon}{c_2}.$$

This *diffracted front* touches the reflected front along the circle determined by the reflexion SML of the critical incident rays.

By (4.8.7) and (4.8.5), $P^{(t)}(N)=0$ so that we may expect the pressure at the diffracted front to be zero. Let us therefore assume that the diffracted pulse behaves near its front like

$$P^{(d)}(c_1t-r\sin\epsilon-(h+x)\cos\epsilon)\,H(c_1t-r\sin\epsilon-(h+x)\cos\epsilon). \quad (4.8.9)$$

Then if E is the point (x, r) on the diffracted front and D the point $(0, r-x\tan\epsilon)$ at the base of the corresponding ray,

$$P^{(d)}(E)=P^{(d)}(D)\left(\frac{r-x\tan\epsilon}{r}\right)^{\frac{1}{2}} \quad (4.8.10)$$

by the intensity law. It remains to determine $P^{(d)}$ on the interface. This can be done by comparing the x-components of acceleration on either side of the interface. We can make this comparison at the point N. In the first medium we must use (4.8.10), whence

$$\left[\frac{\partial p}{\partial x}\right]_N=-\rho_1\left[\frac{\partial u}{\partial t}\right]_N=-P^{(d)}(N)\cos\epsilon.$$

In the second medium

$$\left[\frac{\partial p}{\partial x}\right]_N=-\rho_2\left[\frac{\partial u}{\partial t}\right]_N=+\left[\frac{\partial P^{(r)}}{\partial x}\right]_N.$$

But by (4.8.3), (4.8.5) and (4.8.6) this is the limit of

$$-\frac{1}{\sigma\cos\beta}\frac{1}{R_0}\frac{2\rho_2c_2\cos\alpha}{\rho_2c_2\cos\alpha+\rho_1c_1\cos\beta} \times\left\{\frac{h\sin\epsilon\sec^3\alpha\cos^2\beta}{h\sin\epsilon\sec^3\alpha\cos^2\beta+\sigma}\,\frac{h\tan\alpha}{h\tan\alpha+\sigma\cos\beta}\right\}^{\frac{1}{2}}$$

when $\alpha\to\epsilon$. Then also $\beta\to\frac{1}{2}\pi$, $\sigma\to r_N-h\tan\epsilon$, so that this limit is

$$-\frac{2\tan\epsilon}{r_N^{\frac{1}{2}}(r_N-h\tan\epsilon)^{\frac{3}{2}}}.$$

The two values of the acceleration must be equal; hence

$$P^{(d)}(N)=\frac{2\rho_1}{\rho_2}\frac{\tan\epsilon\sec\epsilon}{r_N^{\frac{1}{2}}(r_N-h\tan\epsilon)^{\frac{3}{2}}}, \quad (4.8.11)$$

and so by (4.8.10)

$$P^{(d)}(E) = \frac{2\rho_1}{\rho_2} \frac{\tan\epsilon \sec\epsilon}{r^{\frac{1}{2}}[r-(x+d)\tan\epsilon]^{\frac{3}{2}}}. \tag{4.8.12}$$

On the 'totally reflected' part of the reflexion front which extends from L in fig. 4.8 to the interface the pressure is probably logarithmically infinite. This is suggested by the known solution for the total reflexion of plane pulses (Howarth, 1948; Friedlander, 1948).

A problem analogous to that of refraction is the reflexion of a spherical pulse by an absorbent plane wall. This has been discussed in detail by Doak (1952).

APPENDIX

The reflexion of a spherical acoustic shock wave

A general formula for the strength of an acoustic shock wave due to the reflexion of a spherical shock at an arbitrary surface has been obtained by J. B. and H. B. Keller (1950). This can be derived in the following way. Let

$$\mathbf{x} = \mathbf{a}(\lambda, \mu) \tag{4.A.1}$$

be the equation of the reflector $\mathscr{R}$ in terms of curvature line parameters λ and μ. Then if $\mathbf{n}$ is the unit vector normal to $\mathscr{R}$ drawn into physical space,

$$\mathrm{d}\mathbf{n} = \frac{\mathbf{a}_\lambda}{f_1}\mathrm{d}\lambda + \frac{\mathbf{a}_\mu}{f_2}\mathrm{d}\mu, \tag{4.A.2}$$

where f_1, f_2 are the principal radii of curvature. Let the source be at $\mathbf{x}_0$ and let $\mathbf{R}$ denote the vector $\mathbf{a}-\mathbf{x}_0$. If the reflected ray at $\mathbf{a}$ is prolonged backwards a distance R then its end-point $\mathbf{X}$ is the image of $\mathbf{x}_0$ on the reflected ray. It will be seen from fig. 4.9 that $\mathbf{x}_0 - \mathbf{X}$ is parallel to $\mathbf{n}$. If the equation of the incident and reflected fronts are respectively $ct = R$ and $ct = \nu$, then the reflected front is given by

$$\mathbf{x} = \mathbf{y} = \mathbf{X} + \nu\mathbf{q}, \tag{4.A.3}$$

where $$\mathbf{X} = 2(\mathbf{R}.\mathbf{n})\,\mathbf{n} = -2R\cos\gamma\,\mathbf{n}, \quad R\mathbf{q} = \mathbf{a} - \mathbf{X}, \tag{4.A.4}$$

γ being the angle between $\mathbf{n}$ and $-\mathbf{R}$.

Let F_1', F_2' be the principal radii of curvature of the surface $\mathbf{x} = \mathbf{X}$. Then the principal radii of curvature of the reflected front at the point $\mathbf{a}$ on $\mathscr{R}$ are $F_1' + R = F_1$, $F_2' + R = F_2$, and the distance σ of the point $\mathbf{y}$ on the reflected ray from **a** is $\nu - R$. Hence by (4.1.7) and (4.1.8) the strength P of the reflected shock is

$$P = \frac{1}{R}\left\{\frac{(F_1' + R)(F_2' + R)}{(F_1' + \nu)(F_2' + \nu)}\right\}^{\frac{1}{2}}, \tag{4.A.5}$$

assuming that the strength of the incident shock is $1/R$. This expression can be put into a different form. Let us denote the curvature line parameters on $\mathbf{x} = \mathbf{X}$ by ρ and σ for the moment. As $\mathbf{q}$ is the unit normal to this surface, we have

$$\mathrm{d}\mathbf{q} = \mathbf{X}_\rho \frac{\mathrm{d}\rho}{F_1'} + \mathbf{X}_\sigma \frac{\mathrm{d}\sigma}{F_2'}.$$

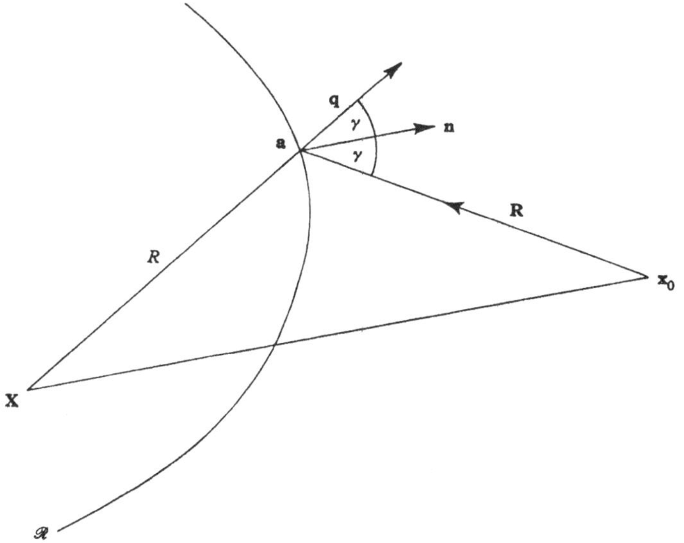

Fig. 4.9. Reflexion of a spherical wave front.

Hence by (4.A.3)

$$\mathrm{d}\mathbf{y} . \mathrm{d}\mathbf{X} = |\mathbf{X}_\rho|^2 \left(1 + \frac{\nu}{F_1'}\right) \mathrm{d}\rho^2 + |\mathbf{X}_\sigma|^2 \left(1 + \frac{\nu}{F_2'}\right) \mathrm{d}\sigma^2 \tag{4.A.6}$$

for fixed ν. On the other hand,

$$\mathrm{d}\mathbf{y} . \mathrm{d}\mathbf{X} = A\,\mathrm{d}\lambda^2 + 2B\,\mathrm{d}\lambda\,\mathrm{d}\mu + C\,\mathrm{d}\mu^2, \tag{4.A.7}$$

where A, B and C are functions of λ, μ and ν. Then by the theory of the invariants of a quadratic differential form,

$$AC - B^2 = |\mathbf{X}_\rho|^2 |\mathbf{X}_\sigma|^2 \left|\frac{\mathrm{D}(\rho, \sigma)}{\mathrm{D}(\lambda, \mu)}\right|^2 \left(1 + \frac{\nu}{F_1'}\right)\left(1 + \frac{\nu}{F_2'}\right).$$

Since only the last two terms on the right-hand side depend on ν we can therefore replace (4.A.5) by

$$P=\frac{1}{R}\left(\frac{A'C'-B'^2}{AC-B^2}\right)^{\frac{1}{2}}, \tag{4.A.8}$$

where A', B' and C' are evaluated on $\mathscr{R}$, that is to say, for $\nu=R$.

We now proceed to calculate A, B and C. By (4.A.3) and (4.A.4),

$$\mathrm{d}\mathbf{y}=\mathrm{d}\mathbf{X}+\nu\,\mathrm{d}\mathbf{q}=\mathrm{d}\mathbf{X}+\nu\left\{\frac{\mathrm{d}\mathbf{a}-\mathrm{d}\mathbf{X}}{R}-\frac{\mathbf{q}}{R}\mathrm{d}R\right\}.$$

Hence since $\mathbf{q}.\mathrm{d}\mathbf{X}=0$,

$$\mathrm{d}\mathbf{y}.\mathrm{d}\mathbf{X}=\left(1-\frac{\nu}{R}\right)|\,\mathrm{d}\mathbf{X}\,|^2+\frac{\nu}{R}\mathrm{d}\mathbf{a}.\mathrm{d}\mathbf{X}. \tag{4.A.9}$$

Again, we have from (4.A.4)

$$\mathrm{d}\mathbf{X}=2(\mathbf{R}.\mathbf{n})\,\mathrm{d}\mathbf{n}+2(\mathbf{R}.\mathrm{d}\mathbf{n})\,\mathbf{n}+2(\mathrm{d}\mathbf{R}.\mathbf{n})\,\mathbf{n}$$

or

$$\mathrm{d}\mathbf{X}=2(\mathbf{R}.\mathbf{n})\,\mathrm{d}\mathbf{n}+2(\mathbf{R}.\mathrm{d}\mathbf{n})\,\mathbf{n}, \tag{4.A.10}$$

since $\mathrm{d}\mathbf{R}=\mathrm{d}\mathbf{a}$ and $\mathbf{n}.\mathrm{d}\mathbf{a}=0$. We substitute this in (4.A.9) and use $\mathbf{n}.\mathrm{d}\mathbf{a}=0$ once more, as well as $\mathbf{n}.\mathrm{d}\mathbf{n}=0$ which holds since $\mathbf{n}$ is a unit vector:

$$\mathrm{d}\mathbf{y}.\mathrm{d}\mathbf{X}=4\left(1-\frac{\nu}{R}\right)\{(\mathbf{R}.\mathbf{n})^2\,|\,\mathrm{d}\mathbf{n}\,|^2+(\mathbf{R}.\mathrm{d}\mathbf{n})^2\}+\frac{2\nu}{R}(\mathbf{R}.\mathbf{n})(\mathrm{d}\mathbf{a}.\mathrm{d}\mathbf{n}). \tag{4.A.11}$$

Finally, it follows from (4.A.2) that

$$|\,\mathrm{d}\mathbf{n}\,|^2=\frac{E}{f_1^2}\mathrm{d}\lambda^2+\frac{G}{f_2^2}\mathrm{d}\mu^2,\quad \mathrm{d}\mathbf{a}.\mathrm{d}\mathbf{n}=\frac{E}{f_1}\mathrm{d}\lambda^2+\frac{G}{f_2}\mathrm{d}\mu^2, \tag{4.A.12}$$

where

$$E=|\,\mathbf{a}_\lambda\,|^2,\quad G=|\,\mathbf{a}_\mu\,|^2, \tag{4.A.13}$$

and that

$$\mathbf{R}.\mathrm{d}\mathbf{n}=\mathbf{R}.\mathbf{a}_\lambda\frac{\mathrm{d}\lambda}{f_1}+\mathbf{R}.\mathbf{a}_\mu\frac{\mathrm{d}\mu}{f_2}.$$

Now $E^{-\frac{1}{2}}\mathbf{a}_\lambda$ and $G^{-\frac{1}{2}}\mathbf{a}_\mu$ are unit vectors in the directions of the curvature lines of R at $\mathbf{a}$. Hence

$$\mathbf{R}.\mathbf{a}_\lambda=-E^{\frac{1}{2}}R\sin\gamma\cos\beta,\quad \mathbf{R}.\mathbf{a}_\mu=-G^{\frac{1}{2}}R\sin\gamma\sin\beta,$$

where β is the angle between the plane containing the incident and reflected rays and the plane defined by $\mathbf{n}$ and $\mathbf{a}_\lambda$. Thus

$$\mathbf{R}.\mathrm{d}\mathbf{n}=-\left\{\frac{E^{\frac{1}{2}}R}{f_1}\cos\beta\,\mathrm{d}\lambda+\frac{G^{\frac{1}{2}}R}{f_2}\sin\beta\,\mathrm{d}\mu\right\}\sin\gamma. \tag{4.A.14}$$

If we now substitute (4.A.12) and (4.A.14) in (4.A.11) and also use the relation $\mathbf{R}.\mathbf{n} = -R\cos\gamma$, then we obtain by comparison with (4.A.7)

$$\tfrac{1}{2}A = 2\left(1-\frac{\nu}{R}\right)\frac{ER^2}{f_1}(\cos^2\gamma+\cos^2\beta\sin^2\gamma) - \frac{\nu E}{f_1}\cos\gamma,$$

$$\tfrac{1}{2}C = 2\left(1-\frac{\nu}{R}\right)\frac{GR^2}{f_2}(\cos^2\gamma+\sin^2\beta\sin^2\gamma) - \frac{\nu G}{f_2}\cos\gamma,$$

$$\tfrac{1}{2}B = 2\left(1-\frac{\nu}{R}\right)\frac{(EG)^{\frac{1}{2}}R^2}{f_1 f_2}\cos\beta\sin\beta\sin^2\gamma.$$

When these values of A, B and C are substituted in (4.A.8) it is found that

$$RP = \left\{\left(\frac{\nu}{R}\right)^2 + 2\nu\left(\frac{\nu}{R}-1\right)\left(2H\cos\gamma + \Gamma\frac{\sin^2\gamma}{\cos\gamma}\right) + 4K(\nu-R)^2\right\}^{-\frac{1}{2}}, \tag{4.A.15}$$

where $$K = \frac{1}{f_1 f_2}, \quad 2H = \frac{1}{f_1}+\frac{1}{f_2}, \quad \Gamma = \frac{\cos^2\beta}{f_1}+\frac{\sin^2\beta}{f_2}. \tag{4.A.16}$$

K is the total (Gaussian) curvature of $\mathscr{R}$ at $\mathbf{a}$, H its mean curvature, and Γ the curvature in the plane which contains the incident and reflected rays (and the normal). The right-hand side of (4.A.15) is the general formula for the divergence factor.

The corresponding formula for an incident plane shock can be derived by making the source recede to infinity. Let us replace $\mathbf{x}_0$, the position vector of the source, by $k\mathbf{b}$, where $\mathbf{b}$ is a unit vector. Then

$$R = k + \mathbf{a}.\mathbf{b} + O(k^{-1})$$

as $\mathbf{k}\to\infty$. We therefore multiply (4.A.15) by k and replace ν by $\nu+k$, and make $k\to\infty$. In the limit, we shall get the strength of the reflected shock due to a plane shock of unit strength incident in the direction opposite to $\mathbf{b}$:

$$P = \left\{1 + 2(\nu-\mathbf{a}.\mathbf{b})\left(2H\cos\gamma+\Gamma\frac{\sin^2\gamma}{\cos\gamma}\right) + 4K(\nu-\mathbf{a}.\mathbf{b})^2\right\}^{-\frac{1}{2}}. \tag{4.A.17}$$

CHAPTER 5

THE DIFFRACTION OF A PULSE BY A WEDGE

1. Introduction

More papers have probably been written about the diffraction of a harmonic wave or a pulse by wedges and half-planes than about any other boundary-value problem for the wave equation. The exact solution (in the harmonic case) was first obtained by Sommerfeld (1896) by applying the method of images on a Riemann surface. The technique of constructing the requisite many-valued solution of the wave equation was simplified by Sommerfeld in a subsequent paper (Sommerfeld, 1897). Finally, Carslaw (1919) replaced the image method by a direct construction of the solution which yields simpler formulae. These had already been obtained independently by Macdonald (1902) by summing the Fourier series representation of the Green's function. The diffraction of pulses was first considered by Sommerfeld (1901) and by Lamb (1910). It was taken up again by Rubinowicz (1927) who derived the Green's function of a wedge in the pulse case. Other methods for obtaining some or all of the principal results both in the harmonic and the pulse case have since been put forward by various authors, including Kay (1953), Herglotz (1952), Garnir (1952 *a*, *b*) (who gives an extensive bibliography), Oberhettinger (1954), Turner (1956) and the author (Friedlander, 1951). The case of an incident plane pulse has been treated by the direct and simple 'cone-field' method by Keller and Blank (1951). Numerical results have been published by the author (Friedlander, 1946 *a* and J. B. Keller, 1952).

Although the literature is extensive it is easily accessible and we shall therefore only give a brief survey of the field. The main result which will be derived is the Green's function of a wedge which is constructed in §3. A Green's function can be defined heuristically as the field due to an instantaneous point source. It satisfies the equation

$$\frac{1}{c^2}\frac{\partial^2 G}{\partial t^2} - \nabla^2 G = \delta(\mathbf{x}-\mathbf{x}_0)\,\delta(t-t_0), \qquad (5.1.1)$$

where $\delta(\mathbf{x}-\mathbf{x}_0)=\delta(x_1-x_1^0)\,\delta(x_2-x_2^0)$ in the two-dimensional and $\delta(x_1-x_1^0)\,\delta(x_2-x_2^0)\,\delta(x_3-x_3^0)$ in the three-dimensional case, δ denoting the Dirac delta function. It must also satisfy the boundary conditions. A precise interpretation of G can be given in terms of the theory of distributions. The theory of Green's functions ('elementary solutions') is of considerable analytical importance. A detailed account would, however, be outside the scope of this book. It is therefore dealt with only briefly in the appendix to this chapter. The method by which the Green's function of the wedge will be constructed is analogous to Macdonald's, but the details are simpler in the pulse case.

2. The Green's function of the wedge

The key to the solution of the wedge diffraction problem is the introduction of certain many-valued solutions of the wave equation. This procedure can be made clear most easily by considering the case where the incident field is specified in terms of a source distribution. Let r, θ be polar coordinates, θ being counted from one face of the wedge. If the angle of the wedge is α then physical space is the domain $r>0$, $0<\theta<\beta$ where $\beta=2\pi-\alpha$. Let the velocity potential Φ satisfy the inhomogeneous wave equation

$$\frac{1}{c^2}\frac{\partial^2\Phi}{\partial t^2}-\nabla^2\Phi=f(r,\theta,t) \tag{5.2.1}$$

in $0<\theta<\beta$, $r>0$. The function f which represents a distribution of sources is to vanish for sufficiently large negative t and in the neighbourhood of the wedge. (This implies the existence of constants T and $\gamma>0$, such that $f=0$ for $t\leqslant T$ and for $r\leqslant\gamma$, $0\leqslant\theta\leqslant\gamma$, $\beta-\gamma\leqslant\theta\leqslant\beta$.) We also assume that the medium is originally undisturbed ($\Phi=0$ for $t\leqslant T$). Taking the wedge to be fixed and rigid, we have the boundary conditions

$$\left[\frac{\partial\Phi}{\partial\theta}\right]_{\theta=0}=0,\quad \left[\frac{\partial\Phi}{\partial\theta}\right]_{\theta=\beta}=0. \tag{5.2.2}$$

It is to be expected that Φ will be singular at the edge $r=0$. Hence a restriction on the behaviour of Φ as $r\to 0$ which ensures the validity of the uniqueness theorem is also required. The proof of the uniqueness theorem depends on the fundamental integral inequality (2.4.2). When applying the procedure of §2.5 to the present case one must first exclude the edge by a small cylinder $r=\epsilon$ in space-time. The theorem will then certainly be valid if the contribution from this cylinder to the integral (2.4.2) vanishes as $\epsilon\to 0$. This will be the case if

$$\frac{\partial\Phi}{\partial t}=O(1), \quad r\frac{\partial\Phi}{\partial r}\to 0 \tag{5.2.3}$$

as $r\to 0$, uniformly in θ for $0\leqslant\theta\leqslant\beta$ and in t for any finite time-interval. This form of the edge condition, which is sufficient but not necessary for the validity of the uniqueness theorem, has a simple physical meaning: the pressure is to remain finite at the edge, and no fluid is to be created or destroyed there.

Let us suppose that f is continuous and that Φ is a strict solution of the wave equation. Since f is only given for $0\leqslant\theta\leqslant\beta$ and vanishes by hypothesis for $\theta=0$, $\theta=\beta$, we can define a continuous periodic function f^* which has the period 2β with respect to θ, coincides with f in $0\leqslant\theta\leqslant\beta$ and is zero in $-\beta\leqslant\theta\leqslant 0$. With this f^* we associate a strict solution of the wave equation,

$$\frac{1}{c^2}\frac{\partial^2\Phi^*}{\partial t^2}-\nabla^2\Phi^*=f^* \tag{5.2.4}$$

that satisfies the edge condition and vanishes for sufficiently large values of $-t$. The uniqueness theorem implies that Φ^* also has period 2β with respect to θ. Then

$$\Phi(r,\theta,t)=\Phi^*(r,\theta,t)+\Phi^*(r,-\theta,t) \tag{5.2.5}$$

satisfies the same conditions as Φ^*, and as it is even in θ it also satisfies the boundary conditions (5.2.2). It is therefore the required velocity potential.

The auxiliary solution Φ^* is a many-valued solution of the wave equation defined for all t on a Riemann surface whose sheets are $(2m-1)\beta\leqslant\theta\leqslant(2m+1)\beta$ $(m=0, \pm 1, \pm 2, \ldots)$. It is uniquely determined on this Riemann surface. Hence we may assume that

$$\Phi^*(r_0,\theta_0,t_0)=E^*(r,\theta,t_0-t;\, r_0,\theta_0)\,.\,f^*(r,\theta,t), \tag{5.2.6}$$

where $E^*(r,\theta,t;\, r_0,\theta_0)$ is a many-valued elementary solution of the wave equation,

$$\frac{1}{c^2}\frac{\partial^2 E^*}{\partial t^2}-\nabla^2 E^*=\frac{1}{r_0}\delta(r-r_0)\,\mathring{\delta}(\theta-\theta_0)\,\delta(t), \tag{5.2.7}$$

which is invariant under a translation 2β in t and whose support is contained in $t \geqslant 0$. Here $\mathring{\delta}(\theta-\theta_0)$ is the delta function with period 2β,

$$\mathring{\delta}(\theta-\theta_0)=\sum_{m=-\infty}^{\infty}\delta(\theta-\theta_0+2m\beta), \tag{5.2.8}$$

and the factor $1/r_0$ has been inserted because the element of area in polar coordinates is $r\,dr\,d\theta$. Hence the solution (5.2.5) of the wedge diffraction problem is

$$\Phi(r_0,\theta_0,t_0)=G(r,\theta,t_0-t;\,r_0,\theta_0)\,.f^*(r,\theta,t), \tag{5.2.9}$$

where

$$G(r,\theta,t;\,r_0,\theta_0)=E^*(r,\theta,t;\,r_0,\theta_0)+E^*(r,\theta,t;\,r_0,\,-\theta_0) \tag{5.2.10}$$

is the Green's function of the wedge. It may be noted that

$$E^*(r,\theta,t;\,r_0,\theta_0)-E^*(r,\theta,t;\,r_0,\,-\theta_0) \tag{5.2.11}$$

is the Green's function of the wedge for the boundary condition $\Phi=0$ $(\theta=0, \theta=\beta)$. The Green's function represents the field due to a point source in the presence of the wedge.

3. Construction of the Green's function

In order to obtain the Green's function we must construct the many-valued elementary solution E^*. This is the field on the Riemann surface due to point sources at $r=r_0$, $\theta=\theta_0+2m\beta$, $t=0$, where $m=0, \pm 1, \ldots$. For $ct<r+r_0$ the disturbance produced by each of these sources will be identical with the ordinary elementary solution of the two-dimensional wave equation, as the dependence domain of a point then meets $t=0$ in a circle which excludes the origin. Let

$$\psi=\theta-\theta_0, \tag{5.3.1}$$

and denote by ψ^* the angle ψ (mod 2β), that is to say, $\psi^*=\psi+2m\beta$, where the integer m is chosen so that $-\beta<\psi^*\leqslant\beta$. Then by (5.A.13)

$$E^*=\frac{c}{2\pi}\frac{\sigma(\psi)\,H(ct-R^*)}{(c^2t^2-R^{*2})^{\frac{1}{2}}} \quad (ct<r+r_0), \tag{5.3.2}$$

where

$$R^*=(r^2+r_0^2-2rr_0\cos\psi^*)^{\frac{1}{2}} \tag{5.3.3}$$

and σ is a periodic function of ψ defined as follows:

$$\sigma(\psi)=1 \quad (|\psi|<\pi); \quad \sigma(\psi)=0 \quad (\pi<|\psi|<\beta); \quad \sigma(\psi+2\beta)=\sigma(\psi). \tag{5.3.4}$$

The crux of the problem is the construction of E^* for $ct \geqslant r+r_0$. An obvious line of attack is the expansion of E^* as a Fourier series in ψ. For

$$\mathring{\delta}(\psi) = \frac{1}{2\beta} + \frac{1}{\beta}\sum_{n=1}^{\infty} \cos \kappa n\psi, \quad \kappa = \frac{\pi}{\beta}. \tag{5.3.5}$$

This equation means that

$$\frac{1}{2\beta}\int_{-\beta}^{\beta} \phi(\theta)\,\mathrm{d}\theta + \frac{1}{\beta}\sum_{n=1}^{N}\int_{-\beta}^{\beta} \phi(\theta)\cos\kappa n(\theta-\theta_0)\,\mathrm{d}\theta \to \phi(\theta_0)$$

holds uniformly in θ_0 as $N \to \infty$ for any test-function $\phi(\theta)$. Test-functions are here indefinitely differentiable periodic functions of θ with period 2β; since the Fourier series of such a function converges uniformly, (5.3.5) is valid.

Let us substitute (5.3.5) in (5.2.7) and assume

$$E^* = \frac{1}{2\beta} U_0(r, t;\, r_0) + \frac{1}{\beta}\sum_{n=1}^{\infty} U_n(r, t;\, r_0) \cos \kappa n\psi. \tag{5.3.6}$$

Then the U_n satisfy the equations

$$\frac{1}{c^2}\frac{\partial^2 U_n}{\partial t^2} - \frac{\partial^2 U_n}{\partial r^2} - \frac{1}{r}\frac{\partial U_n}{\partial r} + \frac{\kappa^2 n^2}{r^2} U_n = \frac{1}{r_0}\delta(r-r_0)\,\delta(t) \quad (n = 0, 1, 2, \ldots), \tag{5.3.7}$$

that is to say, they are the elementary solutions (Riemann functions) of a set of Euler-Darboux equations.

For $ct < r+r_0$, U_n can be calculated directly from (5.3.2):

$$U_n = \frac{c}{2\pi}\int_{\theta_0-\beta}^{\theta_0+\beta} \frac{H(ct-R)\cos\kappa n(\theta-\theta_0)}{(c^2t^2-R^2)^{\frac{1}{2}}}\,\sigma(\theta-\theta_0)\,\mathrm{d}\theta$$

$$= \frac{c}{2\pi}\int_{-\cos^{-1}Z}^{\cos^{-1}Z} \frac{\cos\kappa n\psi}{(2rr_0 Z + 2rr_0\cos\psi)^{\frac{1}{2}}}\,\mathrm{d}\psi,$$

where

$$Z = \frac{c^2t^2 - r^2 - r_0^2}{2rr_0}. \tag{5.3.8}$$

This holds for $-1 \leqslant Z < 1$; when $Z < -1$, $U_n = 0$. Hence

$$U_n = \frac{cH(Z+1)}{\pi\sqrt{(2rr_0)}}\int_0^{\cos^{-1}Z} \frac{\cos\kappa n\psi}{(Z+\cos\psi)^{\frac{1}{2}}}\,\mathrm{d}\psi,$$

or, by a well-known formula†

$$U_n = \begin{cases} \dfrac{c}{2(rr_0)^{\frac{1}{2}}} P_{\kappa n-\frac{1}{2}}(-Z) & (-1 \leqslant Z < 1), \\ 0 & (Z < -1), \end{cases} \tag{5.3.9}$$

where $P_{\kappa n-\frac{1}{2}}$ denotes the Legendre function of the first kind of order $\kappa n - \frac{1}{2}$.

This suggests that we put

$$U_n = (rr_0)^{-\frac{1}{2}} F_n(Z) \quad (Z > 1). \tag{5.3.10}$$

Now U_n satisfies the homogeneous equation (5.3.7) except when $r = r_0$, $t = 0$ and *a fortiori* for $ct > r + r_0$. Thus

$$0 = \left(\frac{1}{c^2}\frac{\partial^2}{\partial t^2} - \frac{\partial^2}{\partial r^2} - \frac{1}{r}\frac{\partial}{\partial r} + \frac{\kappa^2 n^2}{r^2}\right) U_n = \frac{1}{(rr_0)^{\frac{1}{2}}}\left(\frac{1}{c^2}\frac{\partial^2}{\partial t^2} - \frac{\partial^2}{\partial r^2} + \frac{\kappa^2 n^2 - \frac{1}{4}}{r^2}\right) F_n$$

$$= \frac{1}{(rr_0)^{\frac{1}{2}}}\left\{F_n''(Z)\left[\frac{1}{c^2}\left(\frac{\partial Z}{\partial t}\right)^2 - \left(\frac{\partial Z}{\partial r}\right)^2\right] + F_n'(Z)\left(\frac{1}{c^2}\frac{\partial^2 Z}{\partial t^2} - \frac{\partial^2 Z}{\partial r^2}\right) + \frac{\kappa^2 n^2 - \frac{1}{4}}{r^2} F_n\right\}.$$

But

$$\frac{\partial Z}{\partial t} = \frac{c^2 t}{rr_0}, \quad \frac{\partial Z}{\partial r} = \frac{r_0^2 - c^2 t^2}{2r_0 r^2} - \frac{1}{2r_0},$$

$$\frac{\partial^2 Z}{\partial t^2} = \frac{c^2}{rr_0}, \quad \frac{\partial^2 Z}{\partial r^2} = \frac{c^2 t^2 - r_0^2}{r_0 r^3},$$

whence

$$\frac{1}{c^2}\frac{\partial^2 Z}{\partial t^2} - \frac{\partial^2 Z}{\partial r^2} = \frac{r^2 + r_0^2 - c^2 t^2}{r_0 r^3} = -\frac{2Z}{r^2},$$

$$\frac{1}{c^2}\left(\frac{\partial Z}{\partial t}\right)^2 - \left(\frac{\partial Z}{\partial r}\right)^2 = \frac{(r + r_0 + ct)(r - r_0 + ct)(r + r_0 - ct)(r_0 - r + ct)}{2r^4 r_0^2} = \frac{1 - Z^2}{r^2}.$$

Hence F_n must also be a solution of Legendre's equation of order $\kappa n - \frac{1}{2}$,

$$(1 - Z^2) F_n'' - 2Z F_n' + (\kappa^2 n^2 - \tfrac{1}{4}) F_n = 0.$$

Now $Z \to \infty$ as $r_0 \to 0$. In order to be able, ultimately, to satisfy the edge condition we will assume that F_n is to remain finite as

† See, for instance, Hobson (1931), p. 267, eqn. (130).

$Z \to \infty$. It must therefore be a multiple of the Legendre function of the second kind, that is to say,

$$F_n = K\int_{\cosh^{-1}Z}^{\infty} \frac{e^{-\kappa n\xi}}{(\cosh\xi - Z)^{\frac{1}{2}}}\,d\xi \quad (Z > 1), \qquad (5.3.11)$$

where K is a constant (cf. Hobson, 1931, p. 276, eqn. (150)). As $Z \to 1-0$, U_n becomes infinite. This is to be expected since $U_n \cos\kappa n(\theta - \theta_0)$ is a solution of wave equation with a discontinuity on the contracting cylindrical front $ct = r_0 - r$. After passing through the caustic (which is $r = 0$) this is turned into a logarithmic singularity on $ct = r + r_0$ according to the theory developed in the appendix to chapter 3. In fact (Hobson, 1931, p. 225, eqn. (53))

$$U_n \simeq -\frac{c\cos\kappa n\pi}{2\pi(rr_0)^{\frac{1}{2}}}\log(1 - Z),$$

when $1 - Z$ is small and positive. Therefore by (3.A.10)

$$U_n \simeq -\frac{c\cos\kappa n\pi}{2\pi(rr_0)^{\frac{1}{2}}}\log(Z - 1), \qquad (5.3.12)$$

when $Z - 1$ is small and positive. Now we can write (5.3.11)

$$F_n = K\int_{\cosh^{-1}Z}^{\infty} \frac{e^{-\kappa n\xi} - \operatorname{sech}\frac{1}{2}\xi}{(\cosh\xi - Z)^{\frac{1}{2}}}\,d\xi + K\int_{\cosh^{-1}Z}^{\infty} \frac{\operatorname{sech}\frac{1}{2}\xi}{(\cosh\xi - Z)^{\frac{1}{2}}}\,d\xi.$$

As $Z \to 1+0$, the first term tends to a finite constant. The second term can be evaluated by the substitution $\tanh\frac{1}{2}\xi = \tau$; it is

$$\frac{2K}{(Z+1)^{\frac{1}{2}}}\cosh^{-1}\left(\frac{Z+1}{Z-1}\right)^{\frac{1}{2}} = -\frac{K}{\sqrt{2}}\log(Z-1) + O(1).$$

Hence by (5.3.12) $K = (c/\pi\sqrt{2})\cos\kappa n\pi$ and so for $Z > 1$,

$$U_n = \frac{c\cos\kappa n\pi}{\pi(2rr_0)^{\frac{1}{2}}}\int_{\cosh^{-1}Z}^{\infty} \frac{e^{-\kappa n\xi}}{(\cosh\xi - Z)^{\frac{1}{2}}}\,d\xi = \frac{c\cos\kappa n\pi}{\pi(rr_0)^{\frac{1}{2}}}\,Q_{\kappa n - \frac{1}{2}}(Z). \qquad (5.3.13)$$

We can now substitute this in (5.3.6). It is obviously legitimate to invert the order of summation and integration, so that

$$E^* = \frac{c}{\pi\beta(2rr_0)^{\frac{1}{2}}}\int_{\cosh^{-1}Z}^{\infty}\left\{\frac{1}{2} + \sum_{n=1}^{\infty} e^{-\kappa n\xi}\cos\kappa n\pi\,\cos\kappa n\psi\right\}\frac{d\xi}{(\cosh\xi - Z)^{\frac{1}{2}}},$$

whence

$$E^*(r,\theta,t;\,r_0,\theta_0)=\frac{c}{4\pi\beta}\int_{\cosh^{-1}Z}^{\infty}P(\psi,\xi)\frac{\mathrm{d}\xi}{(2rr_0\cosh\xi+r^2+r_0^2-c^2t^2)^{\frac{1}{2}}},\tag{5.3.14}$$

where

$$P(\psi,\xi)=\frac{\sinh\kappa\xi}{\cosh\kappa\xi-\cos\kappa(\psi+\pi)}+\frac{\sinh\kappa\xi}{\cosh\kappa\xi-\cos\kappa(\psi-\pi)}.\tag{5.3.15}$$

The Green's function is thus, for $ct\geqslant r+r_0$,

$$G(r,\theta,t;\,r_0,\theta_0)=\frac{c}{4\pi\beta}\int_{\cosh^{-1}Z}^{\infty}\frac{P(\theta-\theta_0,\xi)+P(\theta+\theta_0,\xi)}{(2rr_0\cosh\xi+r^2+r_0^2-c^2t^2)^{\frac{1}{2}}}\,\mathrm{d}\xi\tag{5.3.16}$$

by (5.2.10). For $ct<r+r_0$ it is given by (5.3.2) and (5.2.10). It can be shown that (5.2.9) then satisfies the edge condition.

4. An alternative form of the Green's function

The integral (5.3.14) can in general only be evaluated numerically and the infinite range of integration is inconvenient. It is therefore desirable to have an alternative form in which the integration is over a finite interval. Such an alternative form can be obtained. It has the additional advantage that it allows one to distinguish between the incident, reflected and diffracted waves.

Consider the contour integral

$$I=\frac{1}{2\pi\mathrm{i}}\int_{\mathscr{C}}\frac{\mathrm{e}^{\kappa\zeta}}{\mathrm{e}^{\kappa\zeta}-\mathrm{e}^{\mathrm{i}\kappa\psi}}\frac{\mathrm{d}\zeta}{(Z+\cosh\zeta)^{\frac{1}{2}}},\tag{5.4.1}$$

where Z is defined by (5.3.8) and $Z>1$, and $\psi=\theta-\theta_0$ as before. The contour $\mathscr{C}$ is a rectangle in the complex ζ-plane with vertices $X\pm\mathrm{i}\pi$, $-X\pm\mathrm{i}\pi$ $(X>1)$, indented at the points $\pm\cosh^{-1}Z\pm\mathrm{i}\pi$ (which are branch points of the integrand) as shown in fig. 5.1. The value of $(Z+\cosh\zeta)^{\frac{1}{2}}$ is fixed by taking it to be real and positive on the real axis. The integrand is single-valued in the interior of $\mathscr{C}$, and so I equals the sum of the residues of the integrand at the poles enclosed by $\mathscr{C}$. These poles are the roots of $\mathrm{e}^{\kappa\zeta}=\mathrm{e}^{\mathrm{i}\kappa\psi}$, and since $\beta=\pi/\kappa$ they are

$$\zeta=\mathrm{i}(\psi+2m\beta)\quad(m=0,\,\pm1,\,\pm2,\,\ldots).$$

Now $\beta>\pi$ by hypothesis. Hence there is just one pole inside $\mathscr{C}$ if $\psi\ (\mathrm{mod}\ 2\beta)$ is between $-\pi$ and π, and no pole at all if $\psi\ (\mathrm{mod}\ 2\beta)$

is between $-\beta$ and $-\pi$ or between π and β. When $\psi \pmod{2\beta} = \pm\pi$ there is a pole on the contour; in this case I will not be defined. Thus we have in terms of (5.3.3) and (5.3.4),

$$I = \frac{(2rr_0)^{\frac{1}{2}}}{\kappa} \frac{\sigma(\psi)}{(c^2t^2 - R^{*2})^{\frac{1}{2}}}. \tag{5.4.2}$$

Thus $\kappa c I/2\pi(2rr_0)^{\frac{1}{2}}$ is a function periodic in θ with period 2β that is equal to the elementary solution of the two-dimensional wave equation in free space when $|\theta-\theta_0| < \pi$, zero when $\pi < |\theta-\theta_0| < \beta$, and defined by its periodicity for all other θ.

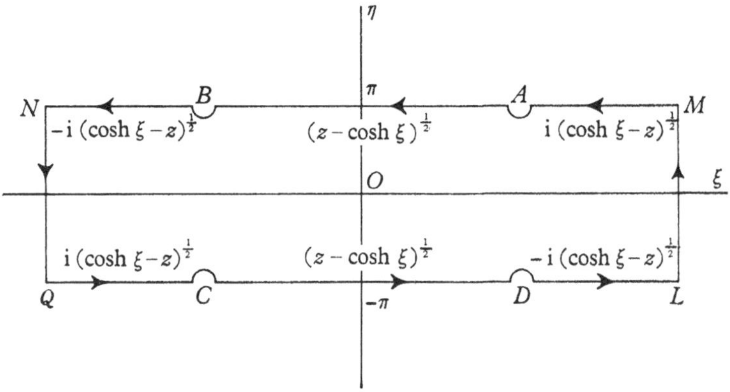

Fig. 5.1. The contour $\mathscr{C}$.

On the other hand, we can evaluate I by making $X \to \infty$ and also letting the radii of the semicircular indentations tend to zero. It is easily verified that the contributions from these and from the lines LM, NQ (fig. 5.1) then tend to zero. Thus I reduces to two integrals along $(-\infty - i\pi, \infty - i\pi)$ and $(-\infty + i\pi, \infty + i\pi)$ respectively. Now writing $\zeta = \xi + i\eta$ we have

$$Z + \cosh\zeta = Z + \cosh\xi \cos\eta + i \sinh\xi \sin\eta.$$

If $|\xi| < \cosh^{-1} Z$ and η varies from 0 to π or from 0 to $-\pi$ for fixed ξ then $\mathscr{R}(Z+\cosh\zeta) > 0$ throughout. Hence $\arg(Z+\cosh\zeta) = 0$ on AB, CD, whence $(Z+\cosh\zeta)^{\frac{1}{2}} = (Z-\cosh\xi)^{\frac{1}{2}}$. If, however, for instance $\xi > \cosh^{-1} Z$ and η increases for fixed ξ from 0 to π, then $\mathscr{R}(Z+\cosh\zeta)$ decreases from $Z+\cosh\xi > 0$ to $Z-\cosh\xi < 0$ while $\mathscr{I}(Z+\cosh\zeta)$ increases from zero to a positive maximum and then decreases to zero again. Hence $\arg(Z+\cosh\zeta)=\pi$ on AM whence $(Z+\cosh\zeta)^{\frac{1}{2}}=\mathrm{i}(\cosh\xi-Z)^{\frac{1}{2}}$. The values of $(Z+\cosh\zeta)^{\frac{1}{2}}$ on DL, BN, QC can be determined similarly and are shown in fig. 5.1. Hence

$$2\pi\mathrm{i}I=\int_{\cosh^{-1}Z}^{\infty}\left\{\frac{e^{\kappa(-\mathrm{i}\pi+\xi)}}{e^{\kappa(-\mathrm{i}\pi+\xi)}-e^{\mathrm{i}\kappa\psi}}+\frac{e^{\kappa(\mathrm{i}\pi+\xi)}}{e^{\kappa(\mathrm{i}\pi+\xi)}-e^{\mathrm{i}\kappa\psi}}\right\}\frac{\mathrm{i}\,d\xi}{(\cosh\xi-Z)^{\frac{1}{2}}}$$
$$+\int_{-\cosh^{-1}Z}^{\cosh^{-1}Z}\left\{\frac{e^{\kappa(-\mathrm{i}\pi+\xi)}}{e^{\kappa(-\mathrm{i}\pi+\xi)}-e^{\mathrm{i}\kappa\psi}}-\frac{e^{\kappa(\mathrm{i}\pi+\xi)}}{e^{\kappa(\mathrm{i}\pi+\xi)}-e^{\mathrm{i}\kappa\psi}}\right\}\frac{d\xi}{(Z-\cosh\xi)^{\frac{1}{2}}}$$
$$+\int_{-\cosh^{-1}Z}^{-\infty}\left\{\frac{e^{\kappa(-\mathrm{i}\pi+\xi)}}{e^{\kappa(-\mathrm{i}\pi+\xi)}-e^{\mathrm{i}\kappa\psi}}+\frac{e^{\kappa(\mathrm{i}\pi+\xi)}}{e^{\kappa(\mathrm{i}\pi+\xi)}-e^{\mathrm{i}\kappa\psi}}\right\}\frac{\mathrm{i}\,d\xi}{(\cosh\xi-Z)^{\frac{1}{2}}}. \tag{5.4.3}$$

The first and the last integrals on the right-hand side are together equal to

$$\int_{\cosh^{-1}Z}^{\infty}\left\{\frac{e^{\kappa(-\mathrm{i}\pi+\xi)}}{e^{\kappa(-\mathrm{i}\pi+\xi)}-e^{\mathrm{i}\kappa\psi}}+\frac{e^{\kappa(\mathrm{i}\pi+\xi)}}{e^{\kappa(\mathrm{i}\pi+\xi)}-e^{\mathrm{i}\kappa\psi}}\right.$$
$$\left.-\frac{e^{\kappa(-\mathrm{i}\pi-\xi)}}{e^{\kappa(-\mathrm{i}\pi-\xi)}-e^{\mathrm{i}\kappa\psi}}-\frac{e^{\kappa(\mathrm{i}\pi-\xi)}}{e^{\kappa(\mathrm{i}\pi-\xi)}-e^{\mathrm{i}\kappa\psi}}\right\}\frac{\mathrm{i}\,d\xi}{(\cosh\xi-Z)^{\frac{1}{2}}}=2\pi\mathrm{i}J,$$

say. Now

$$\frac{e^{\kappa(\pm\mathrm{i}\pi+\xi)}}{e^{\kappa(\pm\mathrm{i}\pi+\xi)}-e^{\mathrm{i}\kappa\psi}}=\frac{1}{2}+\frac{1}{2}\frac{\sinh\kappa\xi+\mathrm{i}\sin\kappa(\psi\pm\pi)}{\cosh\kappa\xi-\cos\kappa(\psi\pm\pi)}, \tag{5.4.4}$$

so that

$$J=\frac{1}{2\pi}\int_{\cosh^{-1}Z}^{\infty}\left\{\frac{\sinh\kappa\xi}{\cosh\kappa\xi-\cos\kappa(\pi+\psi)}+\frac{\sinh\kappa\xi}{\cosh\kappa\xi-\cos\kappa(\pi-\psi)}\right\}\times\frac{d\xi}{(\cosh\xi-Z)^{\frac{1}{2}}},$$

whence by (5.3.14)

$$\frac{cJ}{2\beta\sqrt{(2rr_0)}}=E^*(r,\theta,t;\,r_0,\theta_0). \tag{5.4.5}$$

We have therefore

$$E^*(r,\theta,t;\,r_0,\theta_0)=\frac{c}{2\pi}\frac{\sigma(\psi)}{(c^2t^2-R^{*2})^{\frac{1}{2}}}+\frac{c}{4\pi\beta}\times\int_0^{\cosh^{-1}Z}\frac{Q(\psi,\xi)\,d\xi}{(c^2t^2-r^2-r_0^2-2rr_0\cosh\xi)^{\frac{1}{2}}}, \tag{5.4.6}$$

where by (5.4.3) and (5.4.4)

$$Q(\psi,\xi)=-\frac{\sin\kappa(\pi+\psi)}{\cosh\kappa\xi-\cos\kappa(\pi+\psi)}-\frac{\sin\kappa(\pi-\psi)}{\cosh\kappa\xi-\cos\kappa(\pi-\psi)}, \tag{5.4.7}$$

and if it is understood that the integral in (5.4.6) is to be omitted when $Z < 1$ then that formula represents E^* for $ct < r < r_0$ as well. The Green's function again follows from (5.2.10),

$$\left.\begin{aligned} G(r,\theta,t;\,r_0,\theta_0) &= \frac{c}{2\pi}\frac{\sigma(\theta-\theta_0)}{(c^2t^2-R^{*2})^{\frac{1}{2}}} + \frac{c}{2\pi}\frac{\sigma(\theta+\theta_0)}{(c^2t^2-\bar{R}^{*2})^{\frac{1}{2}}} + G_d, \\ G_d &= \frac{c}{4\pi\beta}\int_0^{\cosh^{-1}Z} \frac{Q(\theta-\theta_0,\xi)+Q(\theta+\theta_0,\xi)}{(c^2t^2-r^2-r_0^2-2rr_0\cosh\xi)^{\frac{1}{2}}}\,d\xi, \end{aligned}\right\} \quad (5.4.8)$$

where
$$\bar{R}^* = \{r^2+r_0^2-2rr_0\cos(\theta+\theta_0)^*\}^{\frac{1}{2}}, \quad (5.4.9)$$

and the asterisk indicates that $\theta+\theta_0$ is to be replaced by $(\theta+\theta_0)$ (mod 2β).

The first term in (5.4.8) is the incident pulse, cut off at the shadow boundary. To see this we note first that by symmetry it is sufficient to take $0 < \theta_0 \leqslant \frac{1}{2}\beta$. Then (since $\pi < \beta \leqslant 2\pi$) $\theta-\theta_0$ increases from $-\theta_0 > -\pi$ to $\beta-\theta_0$ as θ increases from 0 to β, and $\beta-\theta_0 < 2\beta-\pi$ since $\pi-\beta > \frac{1}{2}\beta$. Hence if $0 < \theta_0 < \beta-\pi$ the first term equals the incident field for $0 \leqslant \theta < \theta_0+\pi$ and is zero for $\theta_0+\pi < \theta \leqslant \beta$. When $\pi-\beta \leqslant \theta_0 \leqslant \frac{1}{2}\beta$ there is no shadow and the incident field is always present in physical space. The second term in (5.4.8) is the reflected pulse. The same two cases must again be distinguished. If $0 < \theta_0 \leqslant \beta-\pi$ the incident pulse is reflected by the face $\theta = 0$ of the wedge only. The second term is then $c/2\pi(c^2t^2-\bar{R}^2)^{\frac{1}{2}}$ for $0 \leqslant \theta \leqslant \pi-\theta_0$ and zero for $\pi-\theta_0 < \theta < \beta$. If $\beta-\pi \leqslant \theta_0 < \beta$ then there is a second reflected pulse due to the other face $\theta = \beta$. This differs from zero only in $2\beta-\pi-\theta_0 < \theta < \beta$, and is, by the definition of $\bar{R}^*$,

$$\frac{c}{\{2\pi c^2t^2-r^2-r_0^2+2rr_0\cos(\theta+\theta_0-2\beta)\}^{\frac{1}{2}}},$$

which is the appropriate reflexion of the incident pulse. The third term in (5.4.8) may be called the *diffracted pulse*. It is zero for $ct < r+r_0$. The fronts and the boundaries between the various domains in which incident and reflected pulses occur are shown in fig. 5.2. It should be noted that the sum of these first two terms is in general discontinuous on those of the lines $\theta = \theta_0 \pm \pi$, $\theta = -\theta_0 \pm \pi$ which are in physical space. The diffracted pulse has compensating discontinuities so that G is in fact continuous except at the incident,

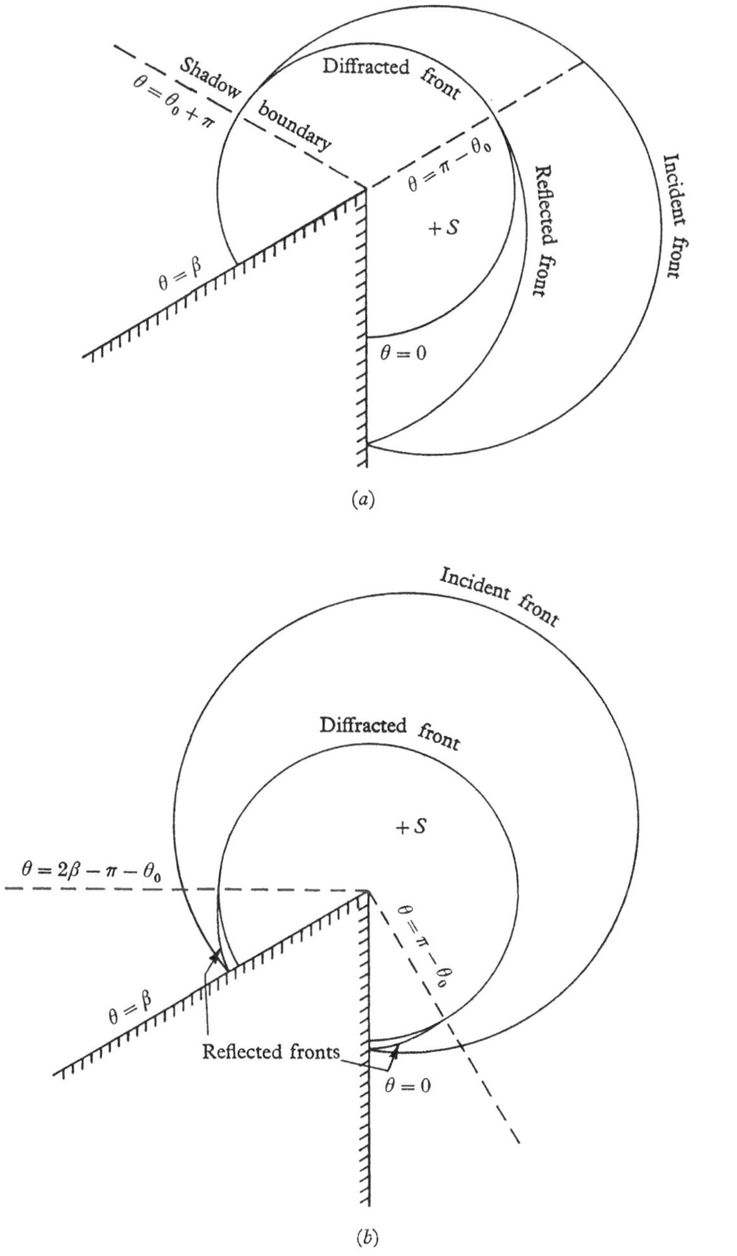

Fig. 5.2. Diffraction of a cylindrical pulse by a wedge (*a*) when a shadow is formed and (*b*) when no shadow is formed. The source is at S.

reflected and diffracted fronts. The derivatives of G behave in the same way.

It is of some interest to consider the diffracted pulse near its front. When $ct-r-r_0$ is small (and positive) then $Z-1$ is also small. Then

$$\frac{c}{4\pi\beta}\int_0^{\cosh^{-1}Z}\frac{Q(\psi,\xi)}{(c^2t^2-r^2-r_0^2+2rr_0\cosh\xi)^{\frac{1}{2}}}\,d\xi$$

$$\simeq\frac{c}{4\pi\beta}\int_0^{\sqrt{\{2(Z-1)\}}}\frac{1}{\sqrt{(2rr_0)}}\frac{d\xi}{(Z-1-\frac{1}{2}\xi^2)^{\frac{1}{2}}}=\frac{c}{4\beta}.$$

Hence by (5.4.8) and (5.4.7)

$$[G_d]_{ct=r+r_0+0}=-\frac{c}{4\beta\sqrt{(rr_0)}}$$

$$\times\left\{\frac{\sin\kappa\pi}{\cos\kappa\pi-\cos\kappa(\theta-\theta_0)}+\frac{\sin\kappa\pi}{\cos\kappa\pi-\cos\kappa(\theta+\theta_0)}\right\}. \quad (5.4.10)$$

Thus the effect of diffraction is to replace the infinity at the incident front by a finite discontinuity at the diffracted front, whose magnitude, however, becomes infinitely large when one of the rays $\theta=\theta_0\pm\pi$ or $\theta=-\theta_0\pm\pi$ is approached.

5. Diffraction of a plane pulse

The field due to an incident plane pulse can be obtained from the Green's function by making the source recede to infinity. If we replace t by $t+r_0/c$, then the incident cylindrical front reaches the edge of the wedge when $t=0$, whatever the position of the source. We can now multiply (5.4.8) by $(2r_0)^{\frac{1}{2}}$ and make $r_0\to\infty$. The first term on the right-hand side then becomes

$$\frac{c}{2\pi}\frac{\sigma(\psi)\,H(ct+r\cos\psi^*)}{(ct+r\cos\psi^*)^{\frac{1}{2}}},$$

so that the incident pulse is in the limit the plane pulse

$$\frac{c}{2\pi}\frac{H\{ct+r\cos(\theta-\theta_0)\}}{\{ct+r\cos(\theta-\theta_0)\}^{\frac{1}{2}}}. \quad (5.5.1)$$

The total field is

$$\lim_{r_0\to\infty}\left\{(2r_0)^{\frac{1}{2}}\,G\left(r,\theta,t+\frac{r_0}{c};\,r_0,\theta_0\right)\right\}$$

$$=\frac{c}{2\pi}\frac{\sigma(\theta-\theta_0)\,H(ct+r\cos(\theta-\theta_0)^*)}{\{ct+r\cos(\theta-\theta_0)^*\}^{\frac{1}{2}}}$$

$$+\frac{c}{2\pi}\frac{\sigma(\theta+\theta_0)\,H(ct+r\cos(\theta+\theta_0)^*)}{\{ct+r\cos(\theta+\theta_0)^*\}^{\frac{1}{2}}}$$

$$+\frac{c}{4\pi\beta}\int_0^{\cosh^{-1}(ct/r)}\frac{Q(\theta-\theta_0,\xi)+Q(\theta+\theta_0,\xi)}{(ct-r\cosh\xi)^{\frac{1}{2}}}\,d\xi. \quad (5.5.2)$$

From this, the field due to an incident plane pulse of arbitrary shape can be deduced by an appeal to the superposition principle. We replace t by $t-t'$, multiply by an arbitrary function $g(t')$, and integrate with respect to t' from $-\infty$ to ∞. Applied to the incident pulse (5.5.1) this procedure yields the incident field

$$\Phi_i=\frac{c}{2\pi}\int_{-\infty}^{t+(r/c)\cos(\theta-\theta_0)}\frac{g(t')\,dt'}{\{c(t-t')+r\cos(\theta-\theta_0)\}^{\frac{1}{2}}}$$

$$=f\left\{t+\frac{r}{c}\cos(\theta-\theta_0)\right\}. \quad (5.5.3)$$

The first two terms on the right-hand side of (5.5.2) give

$$\sigma(\theta-\theta_0)\,f\left\{t+\frac{r}{c}\cos(\theta-\theta_0)^*\right\}+\sigma(\theta+\theta_0)\,f\left\{t+\frac{r}{c}\cos(\theta+\theta_0)^*\right\},$$

while the diffracted field becomes

$$\frac{c}{4\pi\beta}\int_{-\infty}^{t-r/c}g(t')\,dt'\int_0^{\cosh^{-1}c(t-t')/r}\frac{Q(\theta-\theta_0,\xi)+Q(\theta+\theta_0,\xi)}{\{c(t-t')-r\cosh\xi\}^{\frac{1}{2}}}\,d\xi$$

$$=\frac{c}{4\pi\beta}\int_0^{\infty}\{Q(\theta-\theta_0,\xi)+Q(\theta+\theta_0,\xi)\}\,d\xi \times\int_{-\infty}^{t-(r/c)\cosh\xi}\frac{g(t')\,dt'}{\{c(t-t')-r\cosh\xi\}^{\frac{1}{2}}},$$

which is, by (5.5.3),

$$\frac{1}{2\beta}\int_0^{\infty}\{Q(\theta-\theta_0,\xi)+Q(\theta+\theta_0,\xi)\}\,f\left(t-\frac{r}{c}\cosh\xi\right)d\xi.$$

Hence the total field is

$$\Phi = \sigma(\theta-\theta_0) f\left\{t+\frac{r}{c}\cos(\theta-\theta_0)\right\} + \sigma(\theta+\theta_0) f\left\{t+\frac{r}{c}\cos(\theta+\theta_0)^*\right\}$$

$$+\frac{1}{2\beta}\int_0^\infty \{Q(\theta-\theta_0,\xi)+Q(\theta+\theta_0,\xi)\} f\left(t-\frac{r}{c}\cosh\xi\right) d\xi, \quad (5.5.4)$$

where we have replaced $(\theta-\theta_0)^*$ by $\theta-\theta_0$ in the first term, since $|\theta-\theta_0|<\beta$ in physical space.†

A particular case of some interest is that of an incident unit pressure pulse,

$$p_i = \frac{\partial\Phi_i}{\partial t} = H\left\{t+\frac{r}{c}\cos(\theta-\theta_0)\right\}. \quad (5.5.5)$$

The resultant pressure pulse is obviously given by the right-hand side of (5.5.4) with $f(t)=H(t)$. This is the sum of two terms

$$\sigma(\psi) H\left(t+\frac{r}{c}\cos\psi^*\right) + \frac{H(t-r/c)}{2\beta}\int_0^{\cosh^{-1}(ct/r)} Q(\psi,\xi)\, d\xi, \quad (5.5.6)$$

with $\psi=\theta-\theta_0$ and $\psi=\theta+\theta_0$ respectively. Now

$$\frac{1}{2\beta}\int_0^{\cosh^{-1}(ct/r)} Q(\psi,\xi)\, d\xi = -\frac{1}{2\beta}$$

$$\times\int_0^{\cosh^{-1}(ct/r)}\left\{\frac{\sin\kappa(\pi+\psi)}{\cosh\kappa\xi-\cos\kappa(\pi+\psi)}+\frac{\sin\kappa(\pi-\psi)}{\cosh\kappa\xi-\cos\kappa(\pi-\psi)}\right\} d\xi$$

$$=\frac{\sin\kappa\pi}{\beta}\int_0^{\cosh^{-1}(ct/r)}\frac{\cos\kappa\pi-\cos\kappa\psi\cosh\kappa\xi}{(\cosh\kappa\xi-\cos\kappa\pi\cos\kappa\psi)^2-\sin^2\kappa\pi\sin^2\kappa\psi}\, d\xi$$

$$=\frac{1}{\pi}\int_0^{\cosh^{-1}(ct/r)}\frac{v'u-u'v}{u^2+v^2}\, d\xi = \frac{1}{\pi}[\arg(u+iv)]_0^{\cosh^{-1}(ct/r)},$$

where $\quad u=\cosh\kappa\xi\cos\kappa\pi-\cos\kappa\psi, \quad v=\sinh\kappa\xi\sin\kappa\pi.$

Since $\kappa=\pi/\beta$ and $\pi<\beta\leqslant 2\pi$ we have $\frac{1}{2}\leqslant\kappa<1$. Hence $\sin\kappa\pi>0$ and $\cos\kappa\pi\leqslant 0$. Thus v is positive and increases with ξ while u decreases as ξ increases. For $|\psi|>\pi$, $u>0$ when $\xi=0$ so that

$$\frac{1}{\pi}[\arg(u+iv)]_0^{\cosh^{-1}(ct/r)} = \frac{1}{\pi}\left[\tan^{-1}\frac{v}{u}\right]_{\xi=\cosh^{-1}(ct/r)}$$

† It is easily verified that if $F(r,\theta,t)$ satisfies the two-dimensional wave equation then $F(r\sin\gamma,\theta,t+z/c\cos\gamma)$ satisfies the three-dimensional wave equation, r, θ and z being cylindrical polar coordinates. By means of this principle one can at once deduce from (5.5.4) the field due to a plane pulse incident in the direction $(-\sin\gamma\cos\theta_0, -\sin\gamma\sin\theta_0, -\cos\gamma)$ on the wedge.

if $\tan^{-1}$ is taken to be between 0 and π. For $|\psi| < \pi$, $u < 0$ when $\xi = 0$ so that u is negative and

$$\frac{1}{\pi}[\arg(u+iv)]_0^{\cosh^{-1}(ct/r)} = -\frac{1}{\pi}\left[\tan^{-1}\frac{v}{|u|}\right]_{\xi=\cosh^{-1}(ct/r)}.$$

But in this case the first term in (5.5.6), that is to say 1, must be added. Hence the right-hand side of (5.5.6) is again

$$\frac{1}{\pi}\left[\tan^{-1}\frac{v}{u}\right]_{\xi=\cosh^{-1}(ct/r)}.$$

The pressure after the arrival of the diffracted pulse (that is to say, for $ct > r$) is therefore

$$\frac{1}{\pi}\tan^{-1}\left\{\frac{\sinh[\kappa\cosh^{-1}(ct/r)]\sin\kappa\pi}{\cosh\left(\kappa\cosh^{-1}\frac{ct}{r}\right)\cos\kappa\pi-\cos\kappa(\theta-\theta_0)}\right\}$$
$$+\frac{1}{\pi}\tan^{-1}\left\{\frac{\sin\left(\kappa\cosh^{-1}\frac{ct}{r}\right)\sin\kappa\pi}{\cosh\left(\kappa\cosh^{-1}\frac{ct}{r}\right)-\cos\kappa(\theta+\theta_0)}\right\}. \quad (5.5.7)$$

This expression depends only on ct/r and θ. This could have been deduced *a priori* since the wave equation and the boundary conditions as well as the conditions of continuity across the diffracted front are all unchanged when r, t are replaced by kr, kt, where k is a constant.† The solution (5.5.7) can therefore be constructed by means of Busemann's 'cone-field' method which is a familiar technique in the theory of linearized supersonic flow. This is the method of Keller and Blank (1951).

Just behind the diffracted front, $ct-r$ is small. If we take $(ct-r)/r$ to be small then

$$\xi = \cosh^{-1}\frac{ct}{r} \simeq \left(2\frac{ct-r}{r}\right)^{\frac{1}{2}}, \quad \cosh\kappa\xi \simeq 1, \quad \sinh\kappa\xi \simeq \kappa\xi,$$

so that (5.5.7) is approximately equal to

$$\frac{1}{\pi}\tan^{-1}\left\{\frac{\kappa\sin\kappa\pi}{\cos\kappa\pi-\cos\kappa(\theta-\theta_0)}\left(2\frac{ct-r}{r}\right)^{\frac{1}{2}}\right\}$$
$$+\frac{1}{\pi}\tan^{-1}\left\{\frac{\kappa\sin\kappa\pi}{\cos\kappa\pi-\cos\kappa(\theta+\theta_0)}\left(2\frac{ct-r}{r}\right)^{\frac{1}{2}}\right\}. \quad (5.5.8)$$

† The strength of a discontinuity propagated on the diffracted front would be proportional to $r^{-\frac{1}{2}}$, and such a discontinuity would therefore violate the edge condition.

This approximation is valid for all θ. If neither $|\theta-\theta_0|-\pi$ nor $|\theta+\theta_0|-\pi$ is small, then it can be replaced by the simpler one

$$\frac{\kappa\sin\kappa\pi}{\pi}\left\{\frac{1}{\cos\kappa\pi-\cos\kappa(\theta-\theta_0)}+\frac{1}{\cos\kappa\pi-\cos\kappa(\theta+\theta_0)}\right\}\left(2\frac{ct-r}{r}\right)^{\frac{1}{2}}. \tag{5.5.9}$$

When a shadow exists, then the pressure is zero in the shadow for $ct<r$ so that (5.5.7) is then simply the pressure pulse in the shadow. The approximation (5.5.9) shows that the incident acoustic shock is converted into a gradual pressure rise; this is the 'shielding' effect of the wedge which one would expect in a shadow. For a fixed time delay $t-r/c$ this effect increases with r. But as the initial rate of pressure rise is infinite one would expect that in fact a shock wave (weaker than the incident one) is formed.†

6. The half-plane

To illustrate the results which have been obtained, we consider the particularly simple case of the half-plane. This is obtained by putting $\alpha=0$, $\beta=2\pi$, $\kappa=\frac{1}{2}$. Consider first the diffraction of a plane unit pressure pulse,

$$p_i=H\left\{t+\frac{r}{c}\cos(\theta-\theta_0)\right\}. \tag{5.6.1}$$

By symmetry, it is sufficient to take $0\leqslant\theta_0\leqslant\pi$. We must distinguish between three domains which are shown together with the wave fronts in fig. 5.3. Domain I is $0<\theta<\pi-\theta_0$, domain II is $\pi-\theta_0<\theta<\pi+\theta_0$, and domain III is $\pi+\theta_0<\theta<2\pi$. In I, both the incident and the reflected pulse are present; in II, only the incident pulse; III is the shadow.

Until the arrival of the diffracted pulse ($ct<r$) we have in domain I

$$\left.\begin{aligned} p_{\mathrm{I}}&=0, && ct<-r\cos(\theta-\theta_0),\\ p_{\mathrm{I}}&=1, && -r\cos(\theta-\theta_0)<ct<-r\cos(\theta+\theta_0),\\ p_{\mathrm{I}}&=2, && -r\cos(\theta+\theta_0)<ct<r;\end{aligned}\right\} \tag{5.6.2}$$

in domain II

$$\left.\begin{aligned} p_{\mathrm{II}}&=0, && ct<-r\cos(\theta-\theta_0),\\ p_{\mathrm{II}}&=1, && -r\cos(\theta-\theta_0)<ct<r;\end{aligned}\right\} \tag{5.6.3}$$

† The effect of a wedge of angle α near to π on a weak shock wave is investigated in Lighthill (1949).

and in domain III

$$p_{\text{III}} = 0, \quad ct < r. \tag{5.6.4}$$

In order to calculate the pressure for $ct > r$ from (5.5.7) we note first that $\sin\kappa\pi = \sin\frac{1}{2}\pi = 1$, $\cos\kappa\pi = 0$ and

$$\sinh\left(\kappa\cosh^{-1}\frac{ct}{r}\right) = \sinh\left(\tfrac{1}{2}\cosh^{-1}\frac{ct}{r}\right) = \left(\frac{ct-r}{2r}\right)^{\frac{1}{2}}.$$

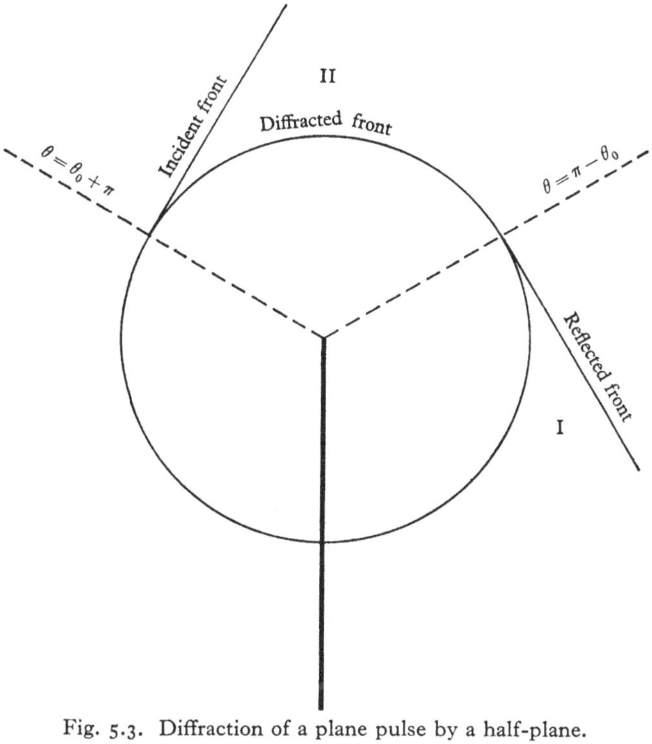

Fig. 5.3. Diffraction of a plane pulse by a half-plane.

Now both $\frac{1}{2}(\theta-\theta_0)$ and $\frac{1}{2}(\theta+\theta_0)$ are in $(-\frac{1}{2}\pi, \frac{1}{2}\pi)$ in domain I. Hence

$$p_{\text{I}} = 2 - \frac{1}{\pi}\tan^{-1}\left\{\sec\frac{\theta-\theta_0}{2}\left(\frac{ct-r}{2r}\right)^{\frac{1}{2}}\right\} - \frac{1}{\pi}\tan^{-1}\left\{\sec\frac{\theta+\theta_0}{2}\left(\frac{ct-r}{2r}\right)^{\frac{1}{2}}\right\}. \tag{5.6.5}$$

In domain II, $\frac{1}{2}(\theta-\theta_0)$ is in $(-\frac{1}{2}\pi, \frac{1}{2}\pi)$ but $\frac{1}{2}(\theta+\theta_0)$ is in $(\frac{1}{2}\pi, \frac{3}{2}\pi)$; hence

$$p_{\text{II}} = 1 - \frac{1}{\pi}\tan^{-1}\left\{\sec\frac{\theta-\theta_0}{2}\left(\frac{ct-r}{2r}\right)^{\frac{1}{2}}\right\} + \frac{1}{\pi}\tan^{-1}\left\{\left|\sec\frac{\theta+\theta_0}{2}\right|\left(\frac{ct-r}{2r}\right)^{\frac{1}{2}}\right\}. \tag{5.6.6}$$

Finally, in the shadow both $\frac{1}{2}(\theta-\theta_0)$ and $\frac{1}{2}(\theta+\theta_0)$ are in $(\frac{1}{2}\pi, \frac{3}{2}\pi)$, so that

$$p_{\text{III}} = \frac{1}{\pi}\tan^{-1}\left\{\left|\sec\frac{\theta-\theta_0}{2}\right|\left(\frac{ct-r}{2r}\right)^{\frac{1}{2}}\right\} + \frac{1}{\pi}\tan^{-1}\left\{\left|\sec\frac{\theta+\theta_0}{2}\right|\left(\frac{ct-r}{2r}\right)^{\frac{1}{2}}\right\}. \tag{5.6.7}$$

The most convenient formula for the calculation of the Green's function is (5.3.2) for $ct < r+r_0$ and (5.3.16) for $ct > r+r_0$. Since $\theta-\theta_0$ increases from $-\theta_0 > -\pi$ to $2\pi-\theta_0 < 2\pi = \beta$ as θ increases from 0 to 2π we have

$$E^* = \begin{cases} \dfrac{c}{2\pi}\dfrac{H(ct-R)}{(c^2t^2-R^2)^{\frac{1}{2}}} & (0 \leqslant \theta < \pi+\theta_0,\ ct < r+r_0), \\ 0 & (\pi+\theta_0 < \theta \leqslant 2\pi,\ ct < r+r_0). \end{cases} \tag{5.6.8}$$

Again, by (5.3.15) $\quad P(\psi, \xi) = \dfrac{2\sinh\xi}{\cosh\xi+\cos\psi}$

in the present case, and so (5.3.14) becomes

$$E^* = \frac{c}{4\pi^2\sqrt{(2rr_0)}}\int_{\cosh^{-1}Z}^{\infty}\frac{\sinh\xi}{\cosh\xi+\cos\psi}\frac{d\xi}{(\cosh\xi-Z)^{\frac{1}{2}}}$$

$$= \frac{c}{4\pi}\frac{1}{(c^2t^2-R^2)^{\frac{1}{2}}} \quad (ct > r+r_0).$$

Taking again $0 < \theta_0 < \pi$ we find the following for the Green's function:

$$G = \begin{cases} \dfrac{c}{2\pi}\dfrac{H(ct-R)}{(c^2t^2-R^2)^{\frac{1}{2}}} + \dfrac{c}{2\pi}\dfrac{H(ct-\bar{R})}{(c^2t^2-\bar{R}^2)^{\frac{1}{2}}} & (0 < \theta < \pi-\theta_0,\ ct < r+r_0), \\ \dfrac{c}{2\pi}\dfrac{H(ct-R)}{(c^2t^2-R^2)^{\frac{1}{2}}} & (\pi-\theta_0 < \theta < \pi+\theta_0,\ ct < r+r_0), \\ 0 & (\pi+\theta_0 < \theta < 2\pi,\ ct < r+r_0), \\ \dfrac{c}{4\pi}\left\{\dfrac{1}{(c^2t^2-R^2)^{\frac{1}{2}}} + \dfrac{1}{(c^2t^2-\bar{R}^2)^{\frac{1}{2}}}\right\} & (0 < \theta < 2\pi,\ ct > r+r_0), \end{cases} \tag{5.6.9}$$

where $\bar{R}$ is the distance of (r, θ) from the image $(r_0, -\theta_0)$ of the source in the plane containing the screen. The simplicity of this result is remarkable.

The Green's function can be used to solve an important problem which is allied to the diffraction problem. Suppose that the normal derivative of a velocity potential Φ (which satisfies the homogeneous wave equation) is given on the half-plane $\theta=0$:

$$\frac{1}{r}\left[\frac{\partial\Phi}{\partial\theta}\right]_{\theta=0}=\frac{1}{r}\left[\frac{\partial\Phi}{\partial\theta}\right]_{\theta=2\pi}=q(r,t). \qquad (5.6.10)$$

Then Φ will be determined uniquely if initial data are given, or an equivalent condition is prescribed. Let us suppose that the medium is originally undisturbed; this is equivalent to saying that the support of Φ is 'compact towards the past', that is to say, it meets *any* dependence domain in a bounded set of points (see the appendix to this chapter). Then Φ can be calculated formally as follows. By definition,

$$\left(\frac{1}{c^2}\frac{\partial^2}{\partial t^2}-\nabla^2\right)G(r,\theta,t_0-t;\,r_0,\theta_0)=\frac{1}{r_0}\delta(r-r_0)\,\delta(\theta-\theta_0)\,\delta(t-t_0).$$

Let us multiply this by $\Phi(r,\theta,t)$ and integrate over physical space, $r>0$, $0<\theta<2\pi$, and all t. The Green's formula of the wave equation gives

$$\begin{aligned}\Phi(r_0,\theta_0,t_0)&=Pf\iiint\Phi\left(\frac{1}{c^2}\frac{\partial^2}{\partial t^2}-\nabla^2\right)Gr\,\mathrm{d}r\,\mathrm{d}\theta\,\mathrm{d}t\\&=Pf\iint\left(\Phi\frac{\partial G}{\partial l}-G\frac{\partial\Phi}{\partial l}\right)\mathrm{d}S, \qquad (5.6.11)\end{aligned}$$

where $\partial/\partial l$ denotes the transversal derivative and the surface integral is over the boundaries $\theta=0$, $\theta=2\pi$. The prefix Pf indicates that the finite part of the divergent integrals in the sense of Hadamard is to be calculated. The contributions from the fronts $c(t_0-t)=R$, $c(t_0-t)=\bar{R}$ are fractional infinities which are ignored; that from the vertex of the dependence domain of (r_0,θ_0,t_0) is taken care of by the delta-function. There is also a contribution from the diffracted front $c(t_0-t)=r_0+r$; when this is examined in detail it is found to consist of line integrals along $\theta=\pi\pm\theta_0$ that cancel (Friedlander, 1951). Now $\partial/\partial l$ on the boundaries is by definition the derivative in the direction of the normal drawn into physical space. Therefore $\partial G/\partial l=0$, and so we are left with

$$\begin{aligned}\Phi(r_0,\theta_0,t_0)=\iint\{&G(r,2\pi,t_0-t;\,r_0,\theta_0)\\&-G(r,0,t_0-t;\,r,\theta_0)\}\,q(r,t)\,\mathrm{d}r\,\mathrm{d}t. \qquad (5.6.12)\end{aligned}$$

By (5.6.9) the Green's function is continuous on the 'screen' $\theta = 0$ for $c(t_0 - t) > r_0 + r$. When $c(t_0 - t) < r_0 + r$ it is equal to the elementary solution doubled by reflexion on the 'illuminated' side of the screen and zero on the other side which is in the shadow. If $\theta_0 < \pi$, the side $\theta = 0$ is illuminated; if $\theta_0 > \pi$ then the other side $\theta = 2\pi$ is illuminated. Hence

$$\left.\begin{aligned} \Phi(r_0, \theta_0, t_0) &= \frac{c \operatorname{sgn}(\theta_0 - \pi)}{\pi} \iint \frac{q(r, t)}{\{c^2(t_0 - t)^2 - R^2\}^{\frac{1}{2}}} \, dr \, dt, \\ R_0 &= (r^2 + r_0^2 - 2rr_0 \cos\theta_0)^{\frac{1}{2}}, \end{aligned}\right\} \tag{5.6.13}$$

the integration being over the domain

$$0 < r < \infty, \quad ct_0 - r_0 - r < ct < ct_0 - R_0. \tag{5.6.14}$$

The formula (5.6.13) will be valid provided that this domain meets the support of $q(r, t)$ in a bounded set. This will certainly be the case if $q = 0$ for $t < \tau(r)$, where

$$\tau'(r) = \frac{dt}{dr} > -\frac{1}{c}. \tag{5.6.15}$$

On the screen the velocity potential is

$$\mp \frac{c}{\pi} \int_0^\infty dr \int_{ct_0 - r_0 - r}^{ct_0 - |r_0 - r|} \frac{q(r, t) \, dt}{\{c^2(t_0 - t)^2 - (r_0 - r)^2\}^{\frac{1}{2}}}, \tag{5.6.16}$$

with the upper sign for $\theta_0 = 0$ and the lower sign for $\theta_0 = 2\pi$. On the prolongation of the screen, $\theta_0 = \pi$, the domain of integration shrinks to the line $c(t_0 - t) = r_0 + r$ and so $\Phi = 0$. This formula is of importance in the theory of linearized supersonic flow.

One can use (5.6.13) to deal with the diffraction of an arbitrary incident field $\Phi_i(r, \theta, t)$ by the half-plane screen $\theta = 0$. For if

$$q = -\frac{1}{r} \left[\frac{\partial \Phi_i}{\partial \theta} \right]_{\theta = 0},$$

then (5.6.13) gives the velocity potential which must be added to the incident one when the screen is present.†

† The condition (5.6.15) then certainly holds in general. For let Φ_i be a pulse solution of the wave equation that vanishes for $t = \tau(r, \theta)$, where τ satisfies the eikonal equation

$$\left(\frac{\partial \tau}{\partial r}\right)^2 + \frac{1}{r^2}\left(\frac{\partial \tau}{\partial \theta}\right)^2 = \frac{1}{c^2}.$$

Then (5.6.15) holds for $t = \tau(r, 0)$ provided that $[\partial\tau/\partial\theta]_{\theta=0} \neq 0$, that is to say, if the incident rays make a finite angle with the screen. The only exceptional case is that of an incident pulse travelling along the screen towards its edge.

If, instead of the normal derivatives, Φ itself is given on the screen, then the Green's function (5.6.9) must be modified by changing the sign of the terms involving $\bar{R}$. The modified Green's function vanishes on the screen, so that the right-hand side of (5.6.11) becomes

$$Pf\iint q'(r,t)\left\{\left[\frac{1}{r}\frac{\partial G}{\partial \theta}\right]_{\theta=0}-\frac{1}{r}\left[\frac{\partial G}{\partial \theta}\right]_{\theta=2\pi}\right\}\mathrm{d}r\,\mathrm{d}t,$$

where $$q'(r,t)=\Phi(r,0,t)=\Phi(r,2\pi,t). \tag{5.6.17}$$

Now $\partial G/r\,\partial\theta$ is zero on that side of the screen which is in the shadow, and double the normal derivative of the elementary solution on the illuminated side. Hence

$$\Phi(r_0,\theta_0,t_0)=-\frac{cr_0\,|\sin\theta_0|}{\pi}Pf\iint\frac{q'(r,t)}{\{c^2(t_0-t)^2-R_0^2\}^{\frac{3}{2}}}\mathrm{d}r\,\mathrm{d}t, \tag{5.6.18}$$

the domain of integration being again defined by (5.6.14). It can be shown that $\Phi\to q'$ when $\theta_0\to 0$, $\theta_0\to 2\pi$. For $\theta_0=\pi$ (5.6.18) must be modified. We have

$$\Phi(r_0,\theta_0,t_0)=-\frac{cr_0\,|\sin\theta_0|}{\pi}Pf\iint\frac{q'(r,t)-q'(r,t_0-R_0/c)}{(c^2(t_0-t)^2-R_0^2)^{\frac{3}{2}}}\mathrm{d}r\,\mathrm{d}t$$
$$+\int_0^\infty q'\left(r,t_0-\frac{R_0}{c}\right)J\,\mathrm{d}r, \tag{5.6.19}$$

where $$J=-\frac{r_0\,|\sin\theta_0|}{\pi}Pf\int_{t_0-(r_0+r/c)}^{t_0-(R_0/c)}\frac{\mathrm{d}t}{\{(t_0-t)^2-R_0^2/c^2\}^{\frac{3}{2}}}.$$

By definition of the finite part of a divergent integral,

$$J=-\frac{r_0\,|\sin\theta_0|}{\pi}\lim_{\epsilon\to 0}\left\{\int_{t_0-(r_0+r)/c}^{t_0-(R_0/c)-\epsilon}\frac{\mathrm{d}t}{\{(t_0-t)^2-R_0^2/c^2\}^{\frac{3}{2}}}+\frac{A}{\epsilon^{\frac{1}{2}}}\right\},$$

A being chosen such that the limit exists. Thus

$$J=-\frac{r_0\,|\sin\theta_0|}{\pi}\lim_{\epsilon\to 0}\left\{\left[\frac{c^2}{R_0^2}\frac{t-t_0}{\{(t_0-t)^2-R_0^2/c^2\}^{\frac{1}{2}}}\right]_{t_0-(r_0+r)/c}^{t_0-(R_0/c)-\epsilon}+\frac{A}{\epsilon^{\frac{1}{2}}}\right\},$$

whence $A=2^{-\frac{1}{2}}(c/R_0)^{\frac{3}{2}}$ and

$$J=\frac{r_0(r_0+r)}{\pi R_0^2}\left(\frac{1-\sin\theta_0}{2rr_0}\right)^{\frac{1}{2}}.$$

Hence $$\lim_{\theta_0\to\pi}J=\frac{1}{\pi}\left(\frac{r_0}{r}\right)^{\frac{1}{2}}\frac{1}{r_0+r}.$$

Now it is easily shown that provided q' is sufficiently regular the first term on the right-hand side of (5.6.19) tends to zero as $\theta_0 \to \pi$, while the passage to the limit under the integral sign in the second term is legitimate. Hence

$$\Phi(r_0, \pi, t_0) = \frac{1}{\pi} \int_0^\infty q'\left(r, t_0 - \frac{r_0 + r}{c}\right) \left(\frac{r_0}{r}\right)^{\frac{1}{2}} \frac{dr}{r_0 + r}, \quad (5.6.20)$$

a formula due to E. N. Fox (Fox, 1948, Appendix B).

7. Some diffraction problems related to the half-plane problem

Once the Green's function of the half-plane is known, the diffraction of a pulse by several non-intersecting half-plane screens can be treated. The secondary field set up consists of pulses generated by each screen, and of the fields which result when these are in turn diffracted by the other screens. By tracing this multiple diffraction process in detail the whole field can be constructed. Another diffraction problem which can be reduced to a combination of half-plane problems is that of the diffraction by a strip with parallel edges. This problem was first solved by E. N. Fox (1948);† in two furthers papers (Fox, 1949, 1951) Fox considered the diffraction of pulses by a slit and by a grating. The propagation of a pulse in the presence of an open-ended semi-infinite channel has been discussed by W. Chester (1950*a*, *b*). We shall not give a detailed account of these investigations here, but consider only briefly how the solutions can be obtained.

The secondary field due to an arbitrary incident field Φ_0 diffracted by a semi-infinite plane screen $\mathscr{S}$ is given by (5.6.13) with $q = -\partial\Phi_0/\partial n$ evaluated on the screen. In this formula, r must now be interpreted as the distance of a variable point on the screen from the edge, R_0, as the distance of this point from the point P at which the field is to be calculated and θ_0 as the angle between the screen and the line from P to the edge. We may write this secondary field symbolically as $-S\Phi_0$, where S is a linear operator. This operator has the following properties: (i) $S\Phi_0$ satisfies the homogeneous

† A similar problem in linearized supersonic aerofoil theory was solved by J. C. Gunn (1947).

wave equation; (ii) $(\partial/\partial n) S\Phi_0 = \partial\Phi_0/\partial n$ on the screen; (iii) the support of $S\Phi_0$ is compact towards the past.†

Suppose now that Φ_0 is incident on two non-intersecting screens $\mathscr{S}_1$ and $\mathscr{S}_2$. With each of these one can associate an operator of the kind we have just defined; let us denote these operators by S_1 and S_2. Consider the velocity potential

$$\Phi = \Phi_0 - S_1\Phi_0 - S_2\Phi_0 + S_1S_2\Phi_0 + S_2S_1\Phi_0 - S_1S_2S_1\Phi_0 - S_2S_1S_2\Phi_0 + \dots. \quad (5.7.1)$$

We note first that for instance $S_1S_2\Phi_0 \neq S_2\Phi_0$, since $S_2\Phi_0$ is not single-valued in the exterior of $\mathscr{S}_1$. Now Φ satisfies the wave equation, and the support of $\Phi - \Phi_0$ is compact towards the past. Also

$$\Phi = \Phi' - S_1\Phi' = \Phi'' - S_2\Phi'',$$

where

$$\Phi' = \Phi_0 - S_2\Phi_0 + S_2S_1\Phi_0 - \dots, \quad \Phi'' = \Phi_0 - S_1\Phi_0 + S_1S_2\Phi_0 + \dots.$$

Hence $\partial\Phi/\partial n = 0$ on both $\mathscr{S}_1$ and $\mathscr{S}_2$. It follows that Φ is the velocity potential of the total field due to the diffraction of the incident pulse by both screens. The series is convergent; it reduces in fact always to a finite sum. For instance, the terms $-S_1\Phi_0$, $S_2S_1\Phi_0$, $-S_1S_2S_1\Phi_0$, ... are generated by the alternate diffraction of the incident pulse at the two screens, beginning with $\mathscr{S}_1$. These pulses obviously arrive at any point one at a time with a finite time delay between each successive pair of pulses. Hence at any instant only a finite number of them differ from zero.

To illustrate (5.7.1), several stages in the diffraction of a plane pulse by a parallel slit are shown in fig. 5.4. In fig. 5.4(*a*) the incident pulse has not reached either edge and the only pulses present are the incident one whose front is I and a reflected pulse with front R. Fig. 5.4(*b*) shows a later stage when the incident front has passed both edges. It is now broken into two segments, one of which has passed through the slit; there is a second reflected pulse with front R_2; finally, there are diffracted pulses with fronts D_1, D_2. A little later one of these, D_1, is diffracted again at the other edge as a pulse with front D_{12}; this stage is shown in fig. 5.4(*c*).

† Property (iii) ensures in general that $S\Phi_0 \neq \Phi_0$. For if Φ_0 is for instance a plane pulse which has the front $ct + x = 0$ so that $\Phi \neq 0$ for $ct + x > 0$ then the dependence domain of any point of $ct > -x$ does *not* meet the support of Φ_0 in a bounded set of points of space-time.

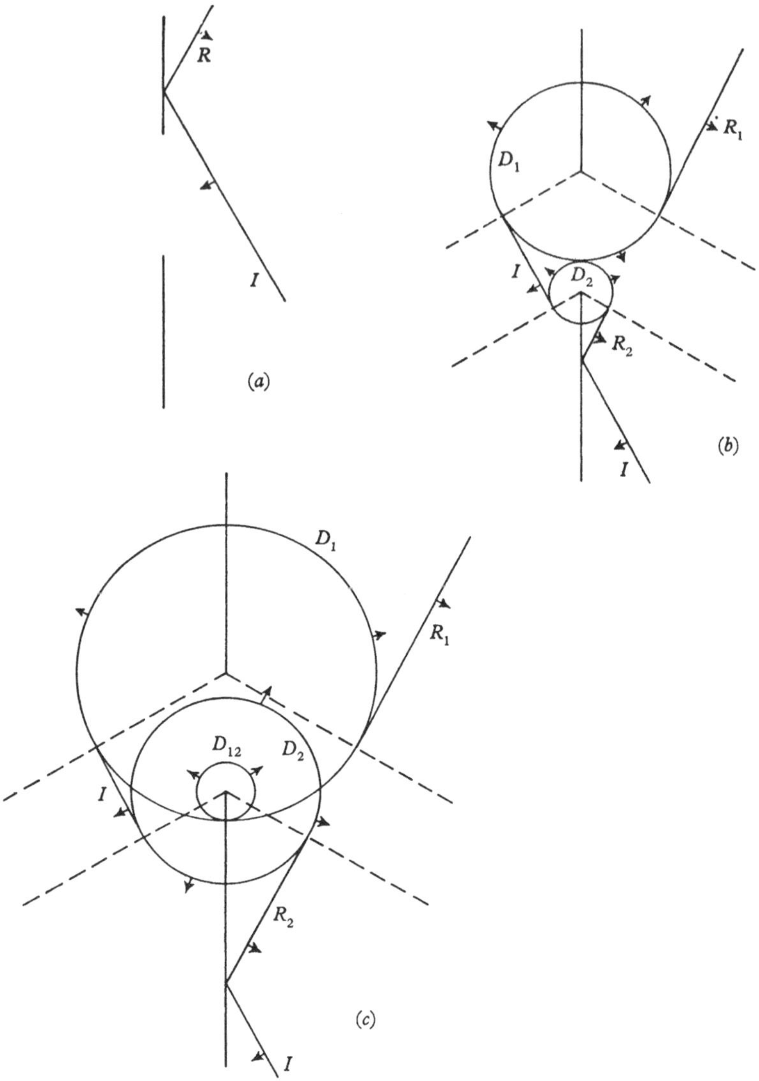

Fig. 5.4. Diffraction of a plane pulse by a parallel slit in a plane screen.

In order to deal with diffraction by a strip we need an important lemma on the diffraction of pulses by a plane screen. Suppose that a plane screen $\mathscr{S}$ which is rigid and fixed occupies some portions of the x-axis and that a pulse Φ_0 is incident on it from $y > 0$. The secondary field Φ_1 is compact towards the past and such that $\partial\Phi_1/\partial y = -\partial\Phi_0/\partial y$ at all points $(x, 0)$ which are on $\mathscr{S}$. Since $-\Phi_1(x, -y, t)$ satisfies the same conditions it follows from the uniqueness theorem that Φ_1 is an odd function of y, continuous on the x-axis at all points not on $\mathscr{S}$ where consequently $\Phi_1 = 0$. These points form the *complementary screen* $\mathscr{S}'$. We can therefore say that the total field $\Phi = \Phi_0 + \Phi_1$ satisfies $\partial\Phi/\partial y = 0$ on $\mathscr{S}$ and $\Phi = \Phi_0$ on $\mathscr{S}'$. Now we can also consider the complementary problem of the diffraction of the same incident pulse by $\mathscr{S}'$ with the 'soft' boundary condition $\Phi' = 0$ on $\mathscr{S}'$. This gives rise to a field $\Phi' = \Phi_0 + \Phi_1'$; we can show by a similar argument that Φ_1' is even and that $\partial\Phi_1'/\partial y = 0$ at all points of $\mathscr{S}$. Hence $\Phi' = 0$ on $\mathscr{S}'$ and $\partial\Phi'/\partial y = \partial\Phi_0/\partial y$ on $\mathscr{S}$. Thus $\Psi = \Phi + \Phi'$ satisfies $\Psi = \Phi_0$ on $\mathscr{S}'$ and $\partial\Psi/\partial y = \partial\Phi_0/\partial y$ on $\mathscr{S}$, and its support is compact towards the past in $y < 0$. Since the incident pulse Φ_0 satisfies the same conditions in $y < 0$ it follows from the uniqueness theorem that

$$\Phi + \Phi' = \Phi_0 \quad (y < 0). \tag{5.7.2}$$

Suppose in particular that the screen is a half-plane $\mathscr{S}$. Then $\Phi = \Phi_0 - S\Phi_0$, where S is the operator already defined. Similarly $\Phi' = \Phi_0 - T'\Phi_0$, where T' is an operator (constructed by means of (5.6.18) and (5.6.20)) such that $T'\Phi_0 = \Phi_0$ on $\mathscr{S}'$, the complementary half-plane, and that $T'\Phi_0$ is compact towards the past. Hence by (5.7.2)

$$S\Phi_0 + T'\Phi_0 = \Phi_0. \tag{5.7.3}$$

We can now consider the diffraction of a pulse by a fixed and rigid strip. Let the strip be $y = 0$, $|x| \leqslant a$. Then by (5.7.2) the diffracted field Φ is equal to $\Phi_0 - \Phi'$, where Φ' is the field due to the diffraction of Φ_0 by a slit occupying $y = 0$, $|x| \leqslant a$ with the soft boundary condition $\Phi' = 0$ on $y = 0$, $|x| > a$. This problem can be treated like that of the slit between rigid planes, so that

$$\Phi' = \Phi_0 - T_1'\Phi_0 - T_2'\Phi_0 + T_1'T_2'\Phi_0 + T_2'T_1'\Phi_0 - \dots,$$

where the operators T_1', T_2' refer to the screens $y = 0$, $x > a$ and $y = 0$, $x < -a$ respectively. Hence

$$\Phi = T_1'\Phi_0 + T_2'\Phi_0 - T_1'T_2'\Phi_0 - T_2'T_1'\Phi_0 + \dots. \tag{5.7.4}$$

This can be put into a different and in practice somewhat more convenient form. Suppose, for instance, that the incident front reaches the edge $y=0$, $x=a$ before it reaches the other edge. Then Φ will at first be identical with the field due to diffraction by the rigid half-plane $y=0$, $x<a$ which can be calculated directly. Now by (5.7.3) $T_1'\Phi_0=\Phi_0-S_1\Phi_0$ in $y<0$, where S_1 is the operator associated with the (hard) screen $y=0$, $x\leqslant a$. But (5.7.4) can be written as

$$\Phi=T_1'\Phi_0+T_2'(\Phi_0-T_1'\Phi_0)-T_1'T_2'(\Phi_0-T_1'\Phi_0)+\dots.$$

Hence $$\Phi=\Phi_0-S_1\Phi_0+T_2'S_1\Phi_0-T_1'T_2'S_1\Phi_0+\dots \qquad (5.7.5)$$

holds in $y<0$. The solution in $y>0$ can be deduced from this since $\Phi-\Phi_0$ is an odd function of y. The meaning of (5.7.5) is easily understood if we consider the disturbances on $y=-0$. The first two terms give $\Phi=\Phi_0$ for $x>a$ and $\partial\Phi/\partial y=0$ for $x<a$. Since we know that $\Phi=\Phi_0$ for $x<-a$, the compensating term $T_2'S_1\Phi_0$ must be introduced next. This in turn is a field which differs from zero for $x>a$ so that a further compensating term $-T_1'T_2'S_1\Phi_0$ is required. Continuing in this way, the whole series (5.7.5) is obtained step by step. The actual evaluation of these terms even on the screen requires laborious calculations; the first few terms have been calculated by Fox (1948) for a plane unit pressure pulse at normal incidence. The diffraction by a mobile strip has been treated by Berry (1952*a*, *b*).

APPENDIX

Elementary solutions and Green's functions

1. *Elementary solutions.* The concept of an elementary solution is of considerable importance in the theory of partial differential equations. Let D be a linear differential operator whose domain consists of functions of n variables $x_1, \dots, x_n$. This operator can also then be applied to distributions associated with test-functions $\phi(x)$. The distribution E is an elementary solution of D if

$$\mathrm{D}(E)=\delta(x-a)=\delta(x_1-a_1)\,\delta(x_2-a_2)\dots\delta(x_n-a_n). \qquad (5.A.1)$$

Since $\delta(x-a).\psi(x)=\psi(a)$, where ψ is a test-function, we have

$$\mathrm{D}(E).\psi(x)=E.\mathrm{D}'(\psi)=\psi(a),$$

where D′ is the differential operator *adjoint* to D. If, for instance,

$$D(\phi)=\sum_{i,j=1}^{n} a_{ij}\frac{\partial^2\phi}{\partial x_i\partial x_j}+\sum_{i=1}^{n} b_i\frac{\partial\phi}{\partial x_i}+c\phi \quad (a_{ij}=a_{ji}),$$

then $$D'(\psi)=\sum_{i,j=1}^{n}\frac{\partial^2}{\partial x_i\partial x_j}(a_{ij}\psi)-\sum_{i=1}^{n}\frac{\partial}{\partial x_i}(b_i\psi)+c\psi.$$

If we put $$D'(\psi)=\phi, \tag{5.A.2}$$

we have therefore $$\psi(a)=E.\phi(x). \tag{5.A.3}$$

Thus an elementary solution allows us to construct a solution of the inhomogeneous adjoint equation when the right-hand side is a test-function. A familiar example in the case $n=3$ and $D=k^2-\nabla^2$ (where k is a real constant) is $\exp\{-k\,|\mathbf{x}-\mathbf{a}|\}/4\pi\,|\mathbf{x}-\mathbf{a}|$. This is a strict solution of the homogeneous equation which is singular at the point $\mathbf{x}=\mathbf{a}$. The elementary solutions of elliptic equations are always of this type. In the hyperbolic case elementary solutions may be weak solutions or distributions. A systematic discussion of the theory of elementary solutions would be out of place in this book. We shall confine ourselves to a brief account of the subject in the case of the wave equation in a homogeneous medium.

2. *Elementary solution of the wave equation in the absence of boundaries.* Consider the inhomogeneous wave equation

$$W(\psi)=\frac{1}{c^2}\frac{\partial^2\psi}{\partial t^2}-\nabla^2\psi=f(\mathbf{x},t), \tag{5.A.4}$$

where c is constant and f is a given function.† It is well known that the *retarded potential*

$$\psi(\mathbf{x},t)=\int f\left\{\mathbf{x}',\,t-\frac{1}{c}\,|\mathbf{x}-\mathbf{x}'|\right\}\frac{d\mathbf{x}'}{4\pi\,|\mathbf{x}-\mathbf{x}'|} \tag{5.A.5}$$

satisfies (5.A.4) provided that f is sufficiently regular, for instance, twice continuously differentiable. Convergence difficulties can be avoided by restricting the source terms f in such a manner that the integral is always over a bounded domain. Since the integration is in effect over the frontier of the dependence domain $\Delta(\mathbf{x},t)$ of $(\mathbf{x},t)$ this can be done by taking f to be a function *compact towards the past* (Leray). By this is meant that the intersection of the support of f, Σ_f, and of any dependence domain, is either compact (closed and bounded) or empty. This implies

† If ψ is the velocity potential then f can be interpreted as the space-time density of sources of fluid; $f\delta x\delta t$ is the volume of fluid produced in the volume δx in the time δt. Such sources are mathematical fictions; they may, however, serve as the mathematical representation of the physical causes of disturbances.

that $f(\mathbf{x}, t) = 0$ for fixed $\mathbf{x}$ when $-t$ is sufficiently large.† The support of ψ is the influence domain of Σ_f which is the union of the influence domains of all the points of Σ_f. It is easily shown that this is also compact towards the past. Hence (5.A.4) is the disturbance produced by the sources f in a medium which was originally undisturbed. We may note that the uniqueness theorem implies that (5.A.4) has at most one solution that is compact towards the past; when it exists it is (5.A.5).

If in (5.A.5) we were to put $f = \delta(\mathbf{x}-\boldsymbol{\xi})\,\delta(t-\tau)$ we would formally obtain the disturbance due to a point source at $(\boldsymbol{\xi}, \tau)$,

$$E'(\boldsymbol{\xi}, \tau \mid \mathbf{x}, t) = \frac{1}{4\pi \mid \mathbf{x}-\boldsymbol{\xi} \mid}\,\delta\left\{t-\tau-\frac{1}{c}\mid \mathbf{x}-\boldsymbol{\xi}\mid\right\}. \tag{5.A.6}$$

This can be justified as follows. If a sequence of distributions T_j is such that $T_j . \phi$ converges as $j \to \infty$ for every test-function ϕ to a limit $T . \phi$, then $T . \phi$ is the functional associated with a distribution T which is called the *weak limit* of the sequence T_j. One then has the important result that the weak limits of the derived sequences $\partial T_j/\partial t$, $\partial T_j/\partial x_\alpha$ are simply the derivatives $\partial T/\partial t$, $\partial T/\partial x_\alpha$ respectively.‡ Now consider a sequence of non-negative source terms f_j which is such that (i) the supports of the f_j tend uniformly to $(\boldsymbol{\xi}, \tau)$ and (ii) $\int f_j \,d\Sigma \to 1$. Then $f_j \to \delta(\mathbf{x}-\boldsymbol{\xi})\,\delta(t-\tau)$ in the weak sense. The retarded potentials ψ_j then tends to $E'(\boldsymbol{\xi}, \tau \mid \mathbf{x}, t)$ in the weak sense. For

$$\begin{aligned}\int \psi_j(\mathbf{x}, t)\,\phi(\mathbf{x}, t)\,d\Sigma &= \int \phi(\mathbf{x}, t)\,d\Sigma \int f_j\left(\mathbf{x}', t-\frac{1}{c}\mid \mathbf{x}-\mathbf{x}'\mid\right)\frac{dx'}{4\pi \mid \mathbf{x}-\mathbf{x}'\mid} \\ &= \int f_j(\mathbf{x}', t')\,d\Sigma' \int \phi\left(\mathbf{x}, t'+\frac{1}{c}\mid \mathbf{x}-\mathbf{x}'\mid\right)\frac{dx}{4\pi \mid \mathbf{x}-\mathbf{x}'\mid} \\ &\to \int \phi\left(\mathbf{x}, \tau+\frac{1}{c}\mid \boldsymbol{\xi}-\mathbf{x}\mid\right)\frac{d\mathbf{x}}{4\pi \mid \mathbf{x}-\boldsymbol{\xi}\mid} \\ &= E'(\boldsymbol{\xi}, \tau \mid \mathbf{x}, t) . \phi(\mathbf{x}, t).\end{aligned}$$

† Since $\Delta(\mathbf{x}, t) \cap \Sigma_f$ is compact there exists a t_1 such that $(\mathbf{x}, t')$ does not belong to it when $t' < t_1$. But $(\mathbf{x}, t') \in \Delta(\mathbf{x}, t)$ when $t' < t$, so $(\mathbf{x}, t')$ cannot belong to Σ_f for $t' < t_1$, that is to say, $f(\mathbf{x}, t') = 0$ for $t' < t_1$. The converse is not true. A sufficient condition for f to be compact towards the past is that $f = 0$ for $t < \sigma(\mathbf{x})$, where $t = \sigma(\mathbf{x})$ is space-like.

‡ The proof is strikingly simple:

$$\frac{\partial T_j}{\partial t} . \phi = -T_j . \frac{\partial \phi}{\partial t} \to -T . \frac{\partial \phi}{\partial t} = \frac{\partial T}{\partial t} . \phi.$$

The assertion that the limit $T . \phi$ of $T_j . \phi$ defines a distribution is, on the other hand, a much deeper result; it follows from the identity of the 'strong' and 'weak' topologies for distributions (Schwartz, 1950/51, I, p. 75).

The commutability of the weak limit and the derivative implies that

$$W(E') = \lim W(\psi_j) = \delta(\mathbf{x}-\boldsymbol{\xi})\,\delta(t-\tau).$$

The distribution E' is therefore an elementary solution of the wave equation.

Since the wave equation is unchanged when t is replaced by $-t$, it is obvious that

$$E(\boldsymbol{\xi},\tau\mid\mathbf{x},t) = \frac{1}{4\pi\mid\mathbf{x}-\boldsymbol{\xi}\mid}\,\delta\left\{\tau - t - \frac{1}{c}\mid\mathbf{x}-\boldsymbol{\xi}\mid\right\} \tag{5.A.7}$$

is also an elementary solution,

$$W(E) = \delta(\mathbf{x}-\boldsymbol{\xi})\,\delta(t-\tau). \tag{5.A.8}$$

Whereas the support of E' is the frontier of the influence domain of $(\boldsymbol{\xi},\tau)$, the support of E is the frontier of the dependence domain of this point. Now (5.A.8) implies that

$$\psi(\boldsymbol{\xi},\tau) = W(E)\,.\,\psi = E\,.\,W(\psi).$$

If we put $W(\psi) = f$ then this becomes the formula of the retarded potential (5.A.5). Obviously ψ is not compact so that this proof is invalid as it stands. But it can be shown that $W(E)\,.\,\psi$, $E\,.\,W(\psi)$ have a meaning if the supports of E and of ψ meet in a compact set. Since the support of E is the frontier of the dependence domain of $(\boldsymbol{\xi},\tau)$ it is therefore sufficient to assume f to be compact towards the past. The proof is then also valid even when f does not satisfy conditions which ensure that the retarded potential is a strict solution of the wave equation, provided that $E\,.f$ has a meaning. In this case, however, the retarded potential is a *weak solution* of the inhomogeneous wave equation.

3. *The uniqueness theorem for distributions that satisfy the wave equation.* Corresponding to the retarded potential one can also define an *advanced potential* which is the unique solution of (5.A.4) that is compact towards the future, provided that f is compact towards the future (that is to say, the intersection of Σ_f and of any influence domain is compact or empty). It is

$$\psi'(\mathbf{x},t) = \int f\left(\mathbf{x}', t + \frac{1}{c}\mid\mathbf{x}-\mathbf{x}'\mid\right)\frac{d\mathbf{x}'}{4\pi\mid\mathbf{x}-\mathbf{x}'\mid}. \tag{5.A.9}$$

The existence of a solution of the wave equation compact towards the future makes it possible to prove a *uniqueness theorem for distributions that satisfy the wave equation.* We note first that if f in (5.A.9) is a test-function (which is compact and therefore compact towards the future) then ψ' is an indefinitely differentiable function that is compact towards the future. Now let T_1, T_2, F_1, F_2 be distributions compact towards the past† such that $W(T_1)=F_1$ and $W(T_2)=F_2$. *Suppose that $F_1=F_2$ in the dependence domain Δ of a fixed point. Then the uniqueness theorem states that $T_1=T_2$ in Δ.* For the compactness towards the past of T_1 and T_2 implies that $W(T_1).\psi'$ and $W(T_2).\psi'$ have a meaning when ψ' is an indefinitely differentiable function compact towards the future. Hence

$$W(T_1).\psi' - W(T_2).\psi' = (F_1 - F_2).\psi' = 0$$

if the support of ψ' is contained in Δ. This equation can also be written as $(T_1-T_2).W(\psi')=0$. But if ϕ is any test-function whose support is contained in Δ, then its advanced potential ψ' has the dependence domain of the support of ϕ for support. This is necessarily also a subset of Δ. Hence $(T_1-T_2).\phi=0$ whenever the support of ϕ is contained in Δ.

If F_1, F_2 are functions and T_1, T_2 are two weak solutions ψ_1, ψ_2 the theorem implies that

$$\int \psi_1 \phi \, d\Sigma = \int \psi_2 \phi \, d\Sigma$$

for all test-functions whose support is contained in Δ. This does not mean that $\psi_1=\psi_2$ in Δ as $\psi_1-\psi_2$ may be a null-function. This indeterminacy is usually of no importance in practice. *A continuous weak solution is uniquely determined.*

4. *The two-dimensional case.* There are corresponding results in the two-dimensional case which can be deduced by Hadamard's *method of descent.* Suppose that f in (5.A.5) depends only on x_1, x_2 and t:

$$\psi = \iiint f\left\{x_1', x_2', t - \frac{1}{c}[(x_1-x_1')^2+(x_2-x_2')^2+(x_3-x_3')^2]^{\frac{1}{2}}\right\} \times \frac{dx_1'\,dx_2'\,dx_3'}{4\pi[(x_1-x_1')^2+(x_2-x_2')^2+(x_3-x_3')^2]^{\frac{1}{2}}}.$$

One can then use

$$t' = t - \frac{1}{c}[(x_1-x_1')^2+(x_1-x_2')^2+(x_3-x_3')^2]^{\frac{1}{2}}$$

instead of x_3' as one of the variables of integration. Now t' decreases from ∞ to $t-\frac{1}{c}|\mathbf{x}-\mathbf{x}'|_2$ as x_3' increases from $-\infty$ to x_3, and then increases again to ∞ as x_3' increases from x_3 to ∞, where

$$|\mathbf{x}-\mathbf{x}'|_2 = \{(x_1-x_1')^2+(x_2-x_2')^2\}^{\frac{1}{2}}. \tag{5.A.10}$$

† Two distributions T and S are said to be equal in an open domain Ω if $T.\phi=S.\phi$ for all test-functions whose supports are contained in Ω; in particular, $T=0$ in Ω if $T.\phi=0$ whenever the support of ϕ is contained in Ω. The support of T is then defined as the complement of the maximal domain in which $T=0$; it is a closed set. Thus $T.\phi=0$ when the intersection of the supports of T and of ϕ is empty.

Hence

$$\psi(x_1, x_2, t) = \frac{c}{2\pi}\iiint \frac{f(x_1', x_2', t')}{\{c^2(t-t')^2 - |\mathbf{x}-\mathbf{x}'|_2^2\}^{\frac{1}{2}}}\,dx_1'\,dx_2'\,dt', \quad (5.\text{A}.11)$$

the integration being over the dependence domain of (x_1, x_2, t),

$$t' \leqslant t - \frac{1}{c}|\mathbf{x}-\mathbf{x}'|_2. \quad (5.\text{A}.12)$$

One can show as in the three-dimensional case that the two-dimensional wave equation has the elementary solutions

$$\left.\begin{aligned} E_2'(\boldsymbol{\xi}, \tau \mid \mathbf{x}, t) &= \frac{c}{2\pi}\frac{H\{c(t-\tau) - |\mathbf{x}-\boldsymbol{\xi}|_2\}}{\{c^2(t-\tau)^2 - |\mathbf{x}-\boldsymbol{\xi}|_2^2\}^{\frac{1}{2}}}, \\ E_2(\boldsymbol{\xi}, \tau \mid \mathbf{x}, t) &= \frac{c}{2\pi}\frac{H\{c(\tau-t) - |\mathbf{x}-\boldsymbol{\xi}|_2\}}{\{c^2(t-\tau)^2 - |\mathbf{x}-\boldsymbol{\xi}|_2^2\}^{\frac{1}{2}}}, \end{aligned}\right\} \quad (5.\text{A}.13)$$

which satisfy

$$W_2(E_2') = W(E_2) = \delta(\mathbf{x}-\boldsymbol{\xi})\,\delta(t-\tau) = \delta(x_1-\xi_1)\,\delta(x_2-\xi_2)\,\delta(t-\tau), \quad (5.\text{A}.14)$$

where

$$W_2 \equiv \frac{1}{c^2}\frac{\partial^2}{\partial t^2} - \frac{\partial^2}{\partial x_1^2} - \frac{\partial^2}{\partial x_2^2}. \quad (5.\text{A}.15)$$

E_2' is the velocity potential in space due to a line source on $x_1 = \xi_1$, $x_2 = \xi_2$ acting at time τ. It is a weak solution of the wave equation except at the line source. As in the three-dimensional case, (5.A.11) is a weak solution of the inhomogeneous wave equation provided the integral converges, and a strict solution if it is twice continuously differentiable.

5. *Laplace transforms.* The Laplace transform of a distribution can be defined.† We can therefore consider the Laplace transforms of the elementary solutions that have been derived. We need here only the following properties of the Laplace transform. (i) If the distribution is a function f that vanishes for fixed $\mathbf{x}$ when $-t$ is sufficiently large then the Laplace transform $\bar{f}$ becomes the usual

$$\bar{f}(\mathbf{x}, s) = \int_{-\infty}^{\infty} e^{-st} f(\mathbf{x}, t)\,dt. \quad (5.\text{A}.16)$$

(ii) Denoting the Laplace transform of T by $\bar{T}$ ($\bar{T}$ is associated with spatial test-functions $\phi(x)$ and depends on the complex parameter s) we have

$$\overline{\frac{\partial T}{\partial x_\alpha}} = \frac{\partial \bar{T}}{\partial x_\alpha}, \quad \overline{\frac{\partial T}{\partial t}} = s\bar{T}. \quad (5.\text{A}.17)$$

(iii) The Laplace transform of $\delta(t)$ is unity, and

$$\overline{\delta(t-\tau)} = e^{-s\tau}, \quad \overline{\delta(\mathbf{x}-\boldsymbol{\xi})\,\delta(t-\tau)} = e^{-s\tau}\,\delta(\mathbf{x}-\boldsymbol{\xi}). \quad (5.\text{A}.18)$$

† Schwartz (1952); Garnir (1952c).

Now it is easy to verify that (5.A.6) can be written as

$$E'(\boldsymbol{\xi}, T \mid \mathbf{x}, t) = \frac{\partial}{\partial t} \frac{H\left\{t-\tau-\frac{1}{c}\,|\,\mathbf{x}-\boldsymbol{\xi}\,|\right\}}{4\pi\,|\,\mathbf{x}-\boldsymbol{\xi}\,|}. \qquad (5.A.19)$$

Hence
$$\bar{E}' = s\int_{\tau+(1/c)\,|\mathbf{x}-\boldsymbol{\xi}|}^{\infty} \frac{e^{-st}}{4\pi\,|\,\mathbf{x}-\boldsymbol{\xi}\,|}\,dt = e^{-s\tau}\frac{e^{-s/c\,|\mathbf{x}-\boldsymbol{\xi}|}}{4\pi\,|\,\mathbf{x}-\boldsymbol{\xi}\,|}. \qquad (5.A.20)$$

In the two-dimensional case (5.A.14) gives

$$\bar{E}_2' = \frac{c}{2\pi}\int_{-\infty}^{\infty} \frac{H(ct-c\tau-|\,\mathbf{x}-\boldsymbol{\xi}\,|_2)}{\{c^2(t-\tau)^2-|\,\mathbf{x}-\boldsymbol{\xi}\,|_2\}^{\frac{1}{2}}}\,dt = \frac{1}{2\pi}\,e^{-s\tau}K_0(s/c\,|\,\mathbf{x}-\boldsymbol{\xi}\,|_2), \qquad (5.A.21)$$

where K_0 is the modified Bessel function of the second kind of order zero. These Laplace transforms are elementary solutions of the corresponding Laplace-transformed wave equations,

$$\frac{s^2}{c^2}\bar{E}' - \nabla^2\bar{E}' = \delta(\mathbf{x}-\boldsymbol{\xi})\,e^{-s\tau}, \qquad (5.A.22)$$

which remain regular as $\mathscr{R}s \to \infty$. This result illustrates a general principle. By (5.A.17) and (5.A.18) the Laplace transforms of E' and E_2' must satisfy (5.A.22) in the sense of the theory of distributions. But as (5.A.22) is elliptic they must therefore be solutions in the ordinary sense, and are in fact the well-known elementary solutions of (5.A.22) in the three- and two-dimensional cases respectively. It may be added that in general the Laplace transform of a weak solution of the wave equation (if it exists) is a strict solution of the Laplace-transformed wave equation.

6. *Reflexion problems; the Green's function.* So far we have considered an unlimited medium. Let us now turn to the case of reflexion by a fixed and rigid obstacle; other boundary conditions can be discussed similarly. The medium then occupies an open domain S of space (physical space) whose frontier $\mathscr{R}$ is the boundary of the reflector. On $\mathscr{R}$ the normal derivative of the velocity potential must vanish. In space-time, S defines a domain Σ that consists of all points $(\mathbf{x}, t)$ such that $\mathbf{x}$ is a point of S; its boundary R is a 3-cylinder parallel to the t-axis which meets $t=0$ in $\mathscr{R}$. Distributions in Σ can be defined by admitting only test-functions whose supports are contained in Σ; they have the same properties as before. But since supports are by definition closed sets of points, and Σ is open, the supports of the test-functions cannot meet R. It is therefore difficult to formulate a boundary-value problem for distributions, and in particular to say what the 'Green's function' of S is. One way to avoid the difficulty is to use Laplace transforms. The results derived in the preceding section suggest that one can define the Green's function as a distribution which is the inverse Laplace transform of the Green's function (in the usual sense) of the Laplace-transformed wave

equation. This method will be used (as a formal device) in the next chapter.†

In order to be able to interpret the results obtained by this indirect method it is, however, helpful to have, as a preliminary, a formal definition of the Green's function and to consider some of its properties. This can be done as follows. Let us postulate the existence of a distribution $G(\mathbf{x}, t \mid \mathbf{x}', t')$ which depends on the coordinates $(\mathbf{x}, t)$ of an arbitrary point as parameters with the following properties. (i) The function

$$\psi(\mathbf{x}, t) = G(\mathbf{x}, t \mid \mathbf{x}', t') \,.\, \phi(\mathbf{x}', t') \tag{5.A.23}$$

(where ϕ is a test-function compact in Σ) is a strict solution of the wave equation $W(\psi) = \phi$ in Σ. (ii) $\partial\psi/\partial n = 0$ on R. (iii) ψ is compact towards the past.‡ Then ψ is uniquely determined and has the influence domain of the support of ϕ as its support. Hence the support of G is contained in the dependence domain of $(\mathbf{x}, t)$. It is also obvious that G is an elementary solution of the wave equation.

We can now formulate the physical concept of the Green's function as the 'disturbance due to a point source'. This results when ϕ tends to a delta function. The crucial point is that the limit must be understood in the weak sense; then ψ also tends to a limit in the weak sense which is a distribution.

In order to establish this we consider first a solution ψ^* of $W(\psi^*) = \phi^*$ (where ϕ^* is another test-function compact in Σ) that satisfies the boundary condition and is compact towards the future. Then $\psi^*(\mathbf{x}, -t)$ is compact towards the past and satisfies

$$W[\psi^*(\mathbf{x}, -t)] = \phi^*(\mathbf{x}, -t)$$

and the boundary condition. Hence by hypothesis

$$\psi^*(\mathbf{x}, -t) = G(\mathbf{x}, t \mid \mathbf{x}', t') \,.\, \phi^*(\mathbf{x}', -t'), \tag{5.A.24}$$

† A rigorous theory on these lines has been developed by J. J. Lions (1955). See also H. G. Garnir (1958). An exposition of this theory would be outside the scope of this monograph.

‡ It is in fact sufficient to assume the existence of a sufficiently regular solution of $W(\psi) = \phi$ that satisfies the boundary condition and is compact towards the past. Then $\psi(\mathbf{x}, t)$ is, for fixed $(\mathbf{x}, t)$, obviously a distributive functional of ϕ. Now one can show that $\psi(\mathbf{x}, t) \to 0$ for fixed $(\mathbf{x}, t)$ if $\phi \to 0$ in the space of test-functions. Suppose, for instance, that the physical space S is unbounded and that from every point of S a semi-cone of semi-angle α ($\alpha > 0$) can be drawn which is contained in S. Then it is not difficult to show that

$$|\psi(\mathbf{x}, t)|^2 \leqslant (48t^2/A) \int (\phi_t^2 + |\nabla\phi|^2) \, d\Sigma, \quad A = 2\pi(1 - \cos\alpha),$$

where the integral is taken over the intersection of the dependence domain $\Delta(\mathbf{x}, t)$ and of Σ. Thus $\psi(\mathbf{x}, t) \to 0$ for fixed $(\mathbf{x}, t)$ if $\phi \to 0$ in the space of test-functions (see footnote ‡, p. 39). It follows that $\psi(\mathbf{x}, t)$ is the functional of a distribution $G(\mathbf{x}, t; \mathbf{x}', t')$ which depends on $(\mathbf{x}, t)$ as parameters, as assumed in the text.

whence $$\psi^*(\mathbf{x}, t) = G(\mathbf{x}, -t \mid \mathbf{x}', -t') \,.\, \phi^*(\mathbf{x}', t'), \tag{5.A.25}$$

where the equivalence of (5.A.24) and (5.A.25) defines the distribution $G(\mathbf{x}, -t \mid \mathbf{x}', -t')$ which is also an elementary solution. Now since ψ is compact towards the past and ψ^* is compact towards the future their supports meet in a compact set. Hence we have the *reciprocity theorem*

$$\psi^* \,.\, \phi = \psi^* \,.\, W(\psi) = W(\psi^*) \,.\, \psi = \psi \,.\, \phi^*. \tag{5.A.26}$$

Substituting (5.A.23) and (5.A.25) we obtain

$$\phi(\mathbf{x}, t) \,.\, \{G(\mathbf{x}, -t \mid \mathbf{x}', -t') \,.\, \phi^*(\mathbf{x}', t')\}$$
$$= \phi^*(\mathbf{x}, t) \,.\, \{G(\mathbf{x}, t \mid \mathbf{x}', t') \,.\, \phi(\mathbf{x}', t')\}. \tag{5.A.27}$$

Now let $\phi_j(\mathbf{x}, t)$ be a sequence of test-functions which tends weakly to $\delta(\mathbf{x})\,\delta(t)$ as $j \to \infty$ and put $\phi = \phi_j(\mathbf{x} - \boldsymbol{\xi}, t - \tau)$. Then $\phi \to \delta(\mathbf{x} - \boldsymbol{\xi})\,\delta(t - \tau)$ as $j \to \infty$. Hence the left-hand side of (5.A.27) tends to a limit in the usual sense,

$$\lim_{j \to \infty} \{G(\mathbf{x}, t \mid \mathbf{x}', t') \,.\, \phi_j(\mathbf{x}' - \boldsymbol{\xi}, t' - \tau)\} \,.\, \phi^*(\mathbf{x}, t) = G(\boldsymbol{\xi}, -\tau \mid \mathbf{x}, -t) \,.\, \phi^*(\mathbf{x}, t).$$

But this equation means that the sequence of functions

$$G(\mathbf{x}, t \mid \mathbf{x}', t') \,.\, \phi_j(\mathbf{x}' - \boldsymbol{\xi}, t' - \tau)$$

of $(\mathbf{x}, t)$ converges weakly to the distribution $G(\boldsymbol{\xi}, -\tau \mid \mathbf{x}, -t)$ as $j \to \infty$. As these functions are also the convolutions of $G(\mathbf{x}, t \mid \mathbf{x}', t')$ and $\phi_j(-\mathbf{x}', -t')$ they converge—because of the continuity of the convolution with respect to either factor—weakly to $G(\mathbf{x}, t \mid \boldsymbol{\xi}, \tau)$. Hence we are left with a reciprocity theorem for the Green's function,

$$G(\mathbf{x}, t \mid \boldsymbol{\xi}, \tau) = G(\boldsymbol{\xi}, -\tau \mid \mathbf{x}, -t), \tag{5.A.28}$$

and have incidentally shown that $G(\mathbf{x}, t \mid \boldsymbol{\xi}, \tau)$ is also a distribution with respect to (x, t).

We can therefore identify $G(\mathbf{x}, t \mid \boldsymbol{\xi}, \tau)$ as the disturbance due to a point source at $(\boldsymbol{\xi}, \tau)$. In particular, let us put

$$G(\mathbf{x}, t \mid \boldsymbol{\xi}, 0) = G(\boldsymbol{\xi}, 0 \mid \mathbf{x}, -t) = G_0(\boldsymbol{\xi}, \mathbf{x}, t) \tag{5.A.29}$$

for the disturbance due to a point source at $(\boldsymbol{\xi}, 0)$. Then $G(\mathbf{x}, t \mid \mathbf{x}', t')$ can be expressed in terms of G_0. This is a consequence of the invariance of both the wave equation and the boundary condition under a translation in t. Thus $\psi(\mathbf{x}, t + h)$ satisfies $W[\psi(\mathbf{x}, t + h)] = \phi(\mathbf{x}, t + h)$ and the boundary condition, and so by (5.A.23)

$$\psi(\mathbf{x}, t + h) = G(\mathbf{x}, t \mid \mathbf{x}', t') \,.\, \phi(\mathbf{x}', t' + h),$$

whence $$\psi(\mathbf{x}, t) = G(\mathbf{x}, t - h \mid \mathbf{x}', t') \,.\, \phi(\mathbf{x}', t' + h).$$

Putting $h=t$ we obtain

$$\psi(\mathbf{x},t)=G(\mathbf{x},0\,|\,\mathbf{x}',t').\phi(\mathbf{x}',t+t')=G(\mathbf{x},0\,|\,\mathbf{x}',-t').\phi(\mathbf{x}',t-t'),$$

and so by (5.A.29)

$$\begin{aligned}\psi(\mathbf{x},t)&=G_0(\mathbf{x},\mathbf{x}',t').\phi(\mathbf{x}',t-t')\\&=G_0(\mathbf{x},\mathbf{x}',t-t').\phi(\mathbf{x}',t')\end{aligned} \tag{5.A.30}$$

(where $G_0(\mathbf{x},\mathbf{x}',t-t')$ is in effect defined by the last equation). We can therefore write

$$G(\mathbf{x},t\,|\,\mathbf{x}',t')=G_0(\mathbf{x},\mathbf{x}',t-t'). \tag{5.A.31}$$

If we substitute this in (5.A.28) we find after putting $t'=0$ that

$$G_0(\mathbf{x},\mathbf{x}',t)=G_0(\mathbf{x}',\mathbf{x},t), \tag{5.A.32}$$

so that G_0 is symmetrical with respect to $\mathbf{x}$ and $\mathbf{x}'$.

The distribution $G_0(\boldsymbol{\xi},\mathbf{x},t)$ will be called the Green's function of S. As we have seen, it is the weak limit of the velocity potential ψ due to the source term ϕ when ϕ tends to $\delta(\mathbf{x}-\boldsymbol{\xi})\,\delta(t)$. According to § 2.6, ψ is the sum of the velocity potential ψ_0 due to the sources ϕ in an unlimited medium and of a secondary disturbance ψ_1 whose front is the reflexion of the front of ψ_0 in the domain of direct reflexion; in the shadow ψ_1 cancels ψ_0 until the arrival of the diffracted front. As ϕ tends to $\delta(\mathbf{x}-\boldsymbol{\xi})\,\delta(t)$, ψ_0 tends obviously to the elementary solution in an unlimited medium with the 'pole' $(\boldsymbol{\xi},0)$. In the two-dimensional case this is, by (5.A.13), the function

$$E_0=\frac{c}{2\pi}\frac{H\{ct-|\,\mathbf{x}-\boldsymbol{\xi}\,|_2\}}{\{c^2t^2-|\,\mathbf{x}-\boldsymbol{\xi}\,|_2^2\}^{\frac{1}{2}}}.$$

Hence G_0-E_0 will also be a function. In the domain of direct reflexion of the rays from $\boldsymbol{\xi}$ this is zero until the arrival of the reflected front, say $ct=\nu$, and becomes infinite like $(ct-\nu)^{-\frac{1}{2}}$ when $ct\to\nu+0$. In the three-dimensional case we have by (5.A.19)

$$E_0=\frac{\partial}{\partial t}\frac{H\left\{t-\frac{1}{c}|\,\mathbf{x}-\boldsymbol{\xi}\,|\right\}}{4\pi\,|\,\mathbf{x}-\boldsymbol{\xi}\,|}.$$

Hence G_0-E_0 is now the derivative of a function with respect to t. This function is again zero for $ct<\nu$ and has a discontinuity on the reflected front whose magnitude can be deduced from the intensity law. In the shadow, G_0 vanishes until the arrival of the diffracted front.

The Laplace transform $\overline{G}_0$ of $G_0(\boldsymbol{\xi},\mathbf{x},t)$ is obviously the Green's function of the Laplace-transformed wave equation,

$$\frac{s^2}{c^2}\overline{G}_0-\nabla^2\overline{G}_0=\delta(\mathbf{x}-\boldsymbol{\xi}). \tag{5.A.33}$$

Its normal derivative vanishes at the boundary, both as a function of $\boldsymbol{\xi}$ and of $\mathbf{x}$.

7. *Solution of the initial-boundary-value problem in terms of the Green's function.* Apart from its intrinsic physical meaning the Green's function is important chiefly because it can be used to construct the solution of the general initial boundary-value problem. Let Φ satisfy the wave equation

$$W(\Phi) = f(\mathbf{x}, t) \tag{5.A.34}$$

in Σ, and suppose that the initial values

$$\Phi_0(\mathbf{x}) = [\Phi]_{t=0+}, \quad \Phi_1(\mathbf{x}) = \left[\frac{\partial \Phi}{\partial t}\right]_{t=0+} \tag{5.A.35}$$

are prescribed in S, as well as

$$\left[\frac{\partial \Phi}{\partial n}\right]_{\mathscr{R}} = q, \tag{5.A.36}$$

where $\partial/\partial n$ denotes the derivative along the normal to $\mathscr{R}$ drawn away from S.

Let us first suppose that Φ is a strict solution of the wave equation. If ψ is a sufficiently regular function and Ω a domain of space-time with frontier F then

$$\int_{\mathscr{R}} \{\Phi W(\psi) - \psi W(\Phi)\}\, \mathrm{d}\Sigma = \int_F \left(\Phi \frac{\partial \psi}{\partial l} - \psi \frac{\partial \Phi}{\partial l}\right) \mathrm{d}S, \tag{5.A.37}$$

where
$$\frac{\partial}{\partial l} = \frac{n_0}{c^2}\frac{\partial}{\partial t} - n_\alpha \frac{\partial}{\partial x_\alpha}$$

is the transversal derivative, (n_1, n_2, n_3, n_0) being the direction cosines of the normal to F drawn away from Ω. This follows from (3.2.3) by omitting the constant factor $1/\rho$. Now let ψ be compact towards the future and satisfy $\partial\psi/\partial n = 0$ on $\mathscr{R}$. We take Ω to be the sub-domain $t > 0$ of Σ. Then the only contributions to the right-hand side of (5.A.37) will be from $t=0$ and from $\mathscr{R}$, since ψ and its derivatives vanish on the boundary of the support of ψ. Since $n_1 = n_2 = n_3 = 0$, $n_0 = -1$ on $t=0$, and $n_0 = 0$ on $\mathscr{R}$ while n_1, n_2, n_3 are the components of the unit normal to $\mathscr{R}$ drawn away from S we find from (5.A.34)–(5.A.36) that

$$\int_S \int_0^\infty \Phi W(\psi)\, \mathrm{d}x\, \mathrm{d}t = \int_S \int_0^\infty f\psi\, \mathrm{d}x\, \mathrm{d}t + \frac{1}{c^2}\int_S \left[\Phi_1 \psi - \Phi_0 \frac{\partial \psi}{\partial t}\right]_{t=0} \mathrm{d}x$$
$$+ \int_{\mathscr{R}} \int_0^\infty \psi q\, \mathrm{d}\mathscr{S}\, \mathrm{d}t. \tag{5.A.38}$$

This equation which is valid when Φ is not necessarily differentiable can be taken as the definition of a 'weak solution' of our initial boundary-value problem.

We can now deduce the value of $\Phi(\mathbf{x}_0, t_0)$ by making $W(\psi)$ tend weakly to $\delta(\mathbf{x} - \mathbf{x}_0)\,\delta(t - t_0)$. We can do this by choosing a sequence of test-functions (compact in Σ) $\phi_j(\mathbf{x}, t)$ that tends weakly to $\delta(\mathbf{x})\,\delta(t)$ and putting $\psi = \psi_j$, where

$$W(\psi_j) = \phi_j(\mathbf{x} - \mathbf{x}_0, t - t_0).$$

Then the left-hand side of (5.A.38) tends weakly to $\Phi(\mathbf{x}_0, t_0)$ provided that $t_0 > 0$, which we assume. Also $\psi_j \to G_0(\mathbf{x}_0, \mathbf{x}, t_0 - t)$, so that the first term on the right-hand side of (5.A.38) tends to

$$G_0(\mathbf{x}_0, \mathbf{x}, t_0 - t) . f(\mathbf{x}, t)\, H(t).$$

The limiting form of the next term can be deduced as follows. It is obvious that ψ_j depends on t and t_0 only as a function of $t - t_0$, so that

$$\left[\frac{\partial \psi_j}{\partial t}\right]_{t=0} = -\left[\frac{\partial \psi_j}{\partial t_0}\right]_{t=0} = -\frac{\partial}{\partial t_0}[\psi_j]_{t=0}.$$

We have therefore

$$\int_S \Phi_1 [\psi_j]_{t=0}\, \mathrm{d}x = -\int_S \int_0^\infty \Phi_1 \frac{\partial \psi_j}{\partial t}\, \mathrm{d}x\, \mathrm{d}t = \frac{\partial}{\partial t_0}\int_S \int_0^\infty \Phi_1 \psi_j\, \mathrm{d}x\, \mathrm{d}t$$

$$-\int_S \Phi_0 \left[\frac{\partial \psi_j}{\partial t}\right]_{t=0} \mathrm{d}x = \int_S \int_0^\infty \Phi_0 \frac{\partial^2}{\partial t^2}[\psi_j]_{t=0}\, \mathrm{d}x\, \mathrm{d}t = \frac{\partial^2}{\partial t_0^2}\int_S \int_0^\infty \Phi_0 \psi_j\, \mathrm{d}x\, \mathrm{d}t.$$

Hence the terms involving Φ_0 and Φ_1 on the right-hand side of (5.A.38) tend to

$$\frac{1}{c^2}\frac{\partial}{\partial t_0}\{G_0(\mathbf{x}_0, \mathbf{x}, t_0 - t) . \Phi_1(\mathbf{x})\, H(t)\} + \frac{1}{c^2}\frac{\partial^2}{\partial t_1^2}\{G_0(\mathbf{x}_0, \mathbf{x}, t_0 - t) . \Phi_0(\mathbf{x})\, H(t)\}.$$

If, finally, we write formally

$$\lim_{j \to \infty} \int_{\mathscr{R}} \int_0^\infty \psi_j q\, \mathrm{d}\mathscr{S}\, \mathrm{d}t = \int \int_0^\infty G_0(\mathbf{x}_0, \mathbf{x}, t_0 - t)\, q(\mathbf{x}, t)\, \mathrm{d}\mathscr{S}\, \mathrm{d}t, \quad (5.\mathrm{A}.39)$$

then we have as the solution of the initial-boundary-value problem

$$\Phi(\mathbf{x}_0, t_0) = G_0(\mathbf{x}_0, \mathbf{x}, t_0 - t) . f(\mathbf{x}, t)\, H(t) + \frac{1}{c^2}\frac{\partial}{\partial t_0}\{G_0(\mathbf{x}_0, \mathbf{x}, t_0 - t) . \Phi_1(\mathbf{x})\, H(t)\}$$
$$+ \frac{1}{c^2}\frac{\partial^2}{\partial t_0^2}\{G_0(\mathbf{x}_0, \mathbf{x}, t_0 - t) . \Phi_0(\mathbf{x})\, H(t)\}$$
$$+ \int_{\mathscr{R}} \int_0^\infty G_0(\mathbf{x}_0, \mathbf{x}, t_0 - t)\, q(\mathbf{x}, t)\, \mathrm{d}\mathscr{S}\, \mathrm{d}t. \quad (5.\mathrm{A}.40)$$

When ct_0 is less than the least distance of $\mathbf{x}_0$ from $\mathscr{R}$, this reduces to the sum of the retarded potential of the sources in $t > 0$ and of Poisson's solution (1.6.2). For then the integral over $\mathscr{R}$ is absent and, according to the remarks made at the end of the preceding section, G_0 reduces to E_0. The term in Φ_1 is

$$\frac{1}{c^2}\frac{\partial}{\partial t_0}\{E_0(\mathbf{x}_0, \mathbf{x}, t_0 - t) . \Phi_1(\mathbf{x})\, H(t)\} = \frac{1}{c^2}\frac{\partial}{\partial t_0}\int_{|\mathbf{x}_0 - \mathbf{x}| \leq ct_0} \frac{\Phi_1(\mathbf{x})\, \mathrm{d}x}{4\pi\, |\mathbf{x}_0 - \mathbf{x}|}.$$

To carry out the differentiation we introduce polar coordinates,

$$x_1 = x_1^0 + r\sin\lambda\cos\mu, \quad x_2 = x_2^0 + r\sin\lambda\sin\mu, \quad x_3 = x_3^0 + r\cos\lambda,$$

and obtain

$$\frac{1}{4\pi c^2}\frac{\partial}{\partial t_0}\int_0^{ct_0}\int_0^{\pi}\int_0^{2\pi}\Phi_1(\mathbf{x})\, r\sin\lambda\, \mathrm{d}r\,\mathrm{d}\lambda\,\mathrm{d}\mu$$

$$= \frac{t_0}{4\pi}\int_0^{\pi}\int_0^{2\pi}\Phi_1(\mathbf{x})\sin\lambda\,\mathrm{d}\lambda\,\mathrm{d}\mu = t_0 M_{ct_0}(\Phi_1).$$

The term involving Φ_0 is therefore

$$\frac{\partial}{\partial t_0}\{t_0 M_{ct_0}(\Phi_0)\}.$$

CHAPTER 6

SOME OTHER DIFFRACTION PROBLEMS

1. Introduction

The relation between geometrical optics and the pulse solutions of the wave equation has been a recurring theme in this book. Since the fronts of the secondary disturbances that arise in scattering problems can be determined by Fermat's principle without solving the wave equation it is natural to inquire how a pulse behaves near its front. A partial answer to this question is given by the theory of geometrical acoustics. But this cannot be applied to diffracted pulse. What happens when a pulse is diffracted by a wedge can be inferred from the explicit solutions developed in the last chapter. However, diffraction by a wedge is essentially an edge effect; it therefore remains to discuss the diffraction of pulses by continuously curved obstacles. There are no general methods of attacking this problem, but in a number of special cases the Laplace transform of the Green's function can be expanded as a series of eigenfunctions which are found by the classical method of separation of variables. Two important examples are the circular cylinder and the sphere.

The solutions obtained are too complicated to be discussed in their entirety, but approximations valid near the reflected and diffracted fronts can be derived from them. The reflected fields are calculated by the method of steepest descent from a contour integral representation of the eigenfunction expansion, and simply confirm the theory of geometrical acoustics. But the approximate evaluation of the diffracted fields yields a new and interesting result. It is found that the diffracted pressure pulse near the diffracted front is always of the form

$$p_d \sim AT^k \exp(-\sigma T^{-\frac{1}{2}}), \tag{6.1.1}$$

where T is the time counted from the arrival of the diffracted front, A and σ are functions of position, and k is a constant; usually k ranges from $-\frac{3}{2}$ to 1. Since the fact that a problem can be solved by the method of separation of variables is a mathematical circumstance of no particular physical significance, it may be conjectured that (6.1.1) is a *general diffraction formula* valid near the diffracted front due to a continuously curved obstacle. There are instances where (6.1.1) is replaced by $AT^k \exp(-\sigma T^{-\mu})$ with $0 < \mu \leqslant 1$.

The diffraction formula (6.1.1) implies that p_d and all its derivatives with respect to the time vanish at the diffracted front. Hence p_d increases at first more slowly than any power of T. Although the diffracted pulse begins at a definite instant its subsequent development is therefore at first more like a diffusion than a propagation process. As a result of this slow build-up of pressure in the shadow the shielding effect of an obstacle in the case of a 'short' pulse is likely to be considerable. This conclusion is, however, only valid in the 'deep' shadow, since the diffraction formula breaks down near the shadow boundary. The magnitude of σ in (6.1.1) is a measure of the depth of a point in the shadow. In the examples which will be considered below, σ is proportional to the distance from the foot of the diffracted ray to the shadow boundary raised to the power $\frac{3}{2}$. Consequently, points in the shadow at some distance from the obstacle are not in the deep shadow in this sense and the pressure may well build up more quickly than the diffraction formula suggests. A satisfactory treatment of the field in the neighbourhood of the shadow boundary has not yet been developed and is one of the unsolved problems of diffraction theory.

From an analytical point of view one can say that p_d has an essential singularity at the diffracted front. This precludes the application of the series expansion method of § 3.9 to diffracted pulses. The inadequacy of this method can be demonstrated quite easily. If we put

$$\left[\frac{\partial^n p_d}{\partial t^n}\right]_{T=0} = P_n,$$

then the P_n satisfy the transport equations associated with the diffracted front. If $P_n = 0$ for $n = 0, 1, \ldots, m-1$ and $P_m \neq 0$, then P_m varies along each diffracted ray according to the intensity law. Now it was shown in § 2.5 that the surface of the reflector in the shadow is a caustic of the diffracted rays. Hence either P_m is finite and not zero in the shadow except on the reflector where it is infinite, or P_m is finite at the reflector but zero at all other points of the shadow. The first alternative is physically untenable. Hence

$P_m = 0$, except perhaps at the reflector. Thus all the P_n, that is to say, all the derivatives $\partial^n p_d/\partial t^n$ vanish at the diffracted front and p_d increases at first more slowly than any power of T. This is of course in agreement with the diffraction formula.

This also follows from the consideration of the diffraction of two-dimensional disturbances by a polygonal cylinder. It has been shown that diffraction by a wedge reduces the frontal infinity of the elementary solution to a finite discontinuity. From this one can deduce that an incident pulse which behaves like $T^k H(T)$ near its front is diffracted as a pulse which increases initially like $T^{k+\frac{1}{2}}H(T)$, where in each case T denotes the time counted from the arrival of the front. Hence an acoustic shock wave diffracted successively at n corners is turned into a pulse which increases at first like $T^{\frac{1}{2}n}$. Since a continuously curved cylinder can be considered as the limit of a polygonal cylinder one is again led to the conclusion that p_d increases more slowly than any power of T in that case.

It would be possible to give a general account of problems to which the method of separation of variables can be applied. But the essential features can be made clear by discussing some representative examples. We shall confine ourselves here to the cases of the circular cylinder and the sphere that have already been mentioned, and to the propagation of a pulse in a stratified medium in convective equilibrium bounded by a rigid plane. The method is borrowed from the theory of propagation of short radio waves round the earth which poses a similar problem.†

2. The Green's function of the circular cylinder

The simplest example of diffraction by a continuously curved obstacle is the scattering of two-dimensional pulses by a circular cylinder (Friedlander, 1954). Taking the axis of the cylinder as z-axis we can work in the xy-plane where it is obviously convenient to introduce polar coordinates r, θ ($x = r\cos\theta$, $y = r\sin\theta$). Let a denote the radius of the cylinder so that its equation is $r = a$.

† See, for instance, Bremmer (1949). The harmonic case involves in effect the calculation of the Fourier transform of the pulse solution, which can be evaluated numerically; in the pulse case considered here the analysis has to be carried one step further so that the requisite approximations for high frequencies—or rather for large values of the Laplace transform parameter s—can be obtained and converted into approximations to the pulse solution.

According to the theory summarized in the appendix to chapter 5 the field due to an arbitrary disturbance can be calculated once the Green's function is known. We shall therefore only consider the Green's function, and the diffraction of plane pulses. The Green's function was defined as the disturbance due to a line source parallel to the cylinder acting at time $t=0$. The general case can obviously be reduced by a rotation to the special one of a source in the xz-plane. Accordingly we shall consider the Green's function $G(x, y, t; r_0)$ which satisfies

$$\frac{1}{c^2}\frac{\partial^2 G}{\partial t^2} - \nabla^2 G = \delta(x-r_0)\,\delta(y)\,\delta(t). \tag{6.2.1}$$

The cylinder is assumed to be rigid and fixed. Since the problem is a two-dimensional one G is a function which satisfies this equation in the weak sense, and the boundary condition

$$\left[\frac{\partial G}{\partial r}\right]_{r=a} = 0, \tag{6.2.2}$$

except when its front (where it becomes infinite) meets the cylinder. Finally, G is compact towards the past; this is equivalent to the initial conditions $G=0$, $\partial G/\partial t=0$ for $t=0$ except at $x=r_0$, $y=0$. G is uniquely determined by these conditions.

As a function of r and θ, G is periodic in θ with period 2π. One could therefore expand G as a Fourier series in θ. But the representation obtained in this way is not suitable for the discussion of the diffracted field. It can be transformed into the appropriate form by Poisson's summation formula, but it is better to approach the problem as follows. Since G is periodic in θ it satisfies for all θ the wave equation (6.2.1) with the right-hand side replaced by the periodic extension of the delta function,

$$\frac{1}{c^2}\frac{\partial^2 G}{\partial t^2} - \left(\frac{\partial^2 G}{\partial r^2} + \frac{1}{r}\frac{\partial G}{\partial r} + \frac{1}{r^2}\frac{\partial^2 G}{\partial \theta^2}\right) = \frac{1}{r}\delta(r-r_0)\,\delta(t)\sum_{m=-\infty}^{\infty}\delta(\theta+2m\pi), \tag{6.2.3}$$

where a factor $1/r$ must be introduced because $\mathrm{d}x\,\mathrm{d}y = r\,\mathrm{d}r\,\mathrm{d}\theta$. This equation can be satisfied by putting

$$G(r,\theta,t;\, r_0) = \sum_{m=-\infty}^{\infty} F(r, \theta+2m\pi, t;\, r_0), \tag{6.2.4}$$

where F is a multiple-valued function of x and y that satisfies

$$\frac{1}{c^2}\frac{\partial^2 F}{\partial t^2}-\left(\frac{\partial^2 F}{\partial r^2}+\frac{1}{r}\frac{\partial F}{\partial r}+\frac{1}{r^2}\frac{\partial^2 F}{\partial \theta^2}\right)=\frac{1}{r}\delta(r-r_0)\,\delta(\theta)\,\delta(t), \quad (6.2.5)$$

and the boundary condition

$$\left[\frac{\partial F}{\partial r}\right]_{r=a}=0. \quad (6.2.6)$$

Like G, F is compact towards the past.

The problem is now reduced to the calculation of F. This will be effected in the next section, but as the result is one of considerable complexity we shall first derive some simple properties of F by geometrical optics. One can interpret F as the Green's function of the cylinder on a Riemann surface $\mathscr{R}$. This has the origin as branch point and its sheets are $(m-1)\pi \leqslant \theta \leqslant (m+1)\pi$ $(m=\ldots-1, 0, 1, \ldots)$; successive sheets are joined along the negative x-axis. The physical plane is the sheet $m=0$. Thus F is the disturbance which results when the elementary solution of the wave equation on $\mathscr{R}$ with the pole $(r_0, 0, 0)$ is scattered by the cylinder $r=a$. There is no need to determine this elementary solution; it is sufficient to observe that for $ct<r_0$ it is necessarily identical with the ordinary elementary solution

$$E(r,\theta,t;\,r_0)=\frac{c}{2\pi}\frac{H(ct-R)}{(c^2t^2-R^2)^{\frac{1}{2}}}, \quad R=(r^2+r_0^2-2rr_0\cos\theta)^{\frac{1}{2}}. \quad (6.2.7)$$

This is a consequence of the uniqueness theorems which can be extended to solutions of the wave equation defined on $\mathscr{R}$. We must therefore consider the reflected and diffracted fronts on $\mathscr{R}$ produced by the incident front $ct=R$. The incident rays are the straight lines issuing from the source point $S(r_0, 0)$. An incident ray inclined to the negative x-axis at an angle less than $\cos^{-1}(a/r_0)$† meets the cylinder and is reflected. The two glancing rays SA, SB touch the circle $r=a$ at the points A $(a, \cos^{-1}(a/r_0))$ and B $(a, -\cos^{-1}(a/r_0))$ (fig. 6.1). Their prolongations AA' and BB' are the shadow boundaries. The incident front reaches A and B simultaneously at time $(r_0^2-a^2)^{\frac{1}{2}}/c$, and at a later instant t extends from a point A' on one shadow boundary to a point B' on the other shadow boundary. At A' and B' it touches the reflected front. Both the incident and

† It will be understood throughout that $O \leqslant \cos^{-1} z \leqslant 1$.

the reflected fronts are confined to the sheet $m=0$. There are obviously two diffracted fronts which join the incident front at A' and B' respectively. The other end-point C of the first of these travels round the circle $r=a$ with velocity c in the positive sense; hence this front propagates into $\theta>0$. The second diffracted front

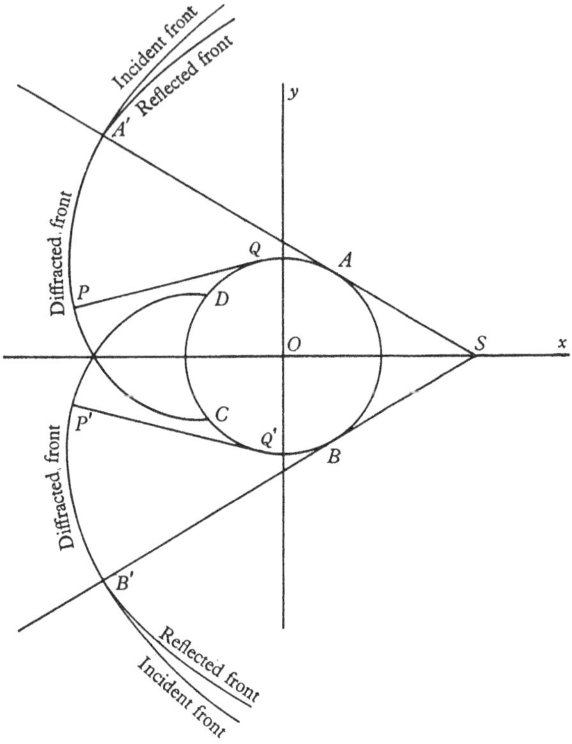

Fig. 6.1. Reflexion and diffraction of a cylindrical wave front by a circular cylinder.

$B'D$ propagates into $\theta<0$; its end-point D travels round the circle in the negative sense with velocity c. The shadow on $\mathscr{R}$ consists of the two domains swept out by these two fronts, and the arrival time τ of the diffracted front at any point of the shadow is a single-valued function of position on $\mathscr{R}$.

It is not difficult to calculate τ. Let first $P(r,\theta)$ be a point in the shadow in $\theta>0$. The 'Fermat path' from the source S to P consists of the glancing ray SA, the circular arc AQ and the diffracted ray

QP which touches the circle at Q. The length of this Fermat path is $c\tau$; if we denote the angle QOA by δ then

$$c\tau=(r_0^2-a^2)^{\frac{1}{2}}+(r^2-a^2)^{\frac{1}{2}}+a\delta. \qquad (6.2.8)$$

The same formula holds for a point P' in $\theta<0$. The Fermat path is now $SBQ'P'$, and δ is defined as the magnitude of the angle $Q'OB$. To determine δ we observe that SA, SB each subtend an angle $\cos^{-1}(a/r_0)$ at O, while PQ, $P'Q'$ subtend in each case the angle $\cos^{-1}(a/r)$. In the case of P we have

$$\theta=\cos^{-1}\frac{a}{r_0}+\delta+\cos^{-1}\frac{a}{r},$$

while for P'
$$\theta=-\cos^{-1}\frac{a}{r_0}-\delta-\cos^{-1}\frac{a}{r}.$$

Hence both cases are covered by

$$\delta=|\theta|-\cos^{-1}\frac{a}{r_0}-\cos^{-1}\frac{a}{r}. \qquad (6.2.9)$$

The inequality $\delta>0$ defines the shadow and $\delta=0$ is the joint equation of the two shadow boundaries. In the domain of direct reflexion δ is negative.

We can now state that in the shadow $F(r,\theta,t,r_0)=0$ for $t<\tau$. In the domain of direct reflexion, $F=0$ for $ct<R$. Hence the series (6.2.4) is in fact a finite sum. Its terms can be thought of as diffracted pulses that have encircled the cylinder a number of times in one sense or the other. It is an obvious consequence of the symmetry of the problem that G and F are even functions of θ. It is therefore sufficient to consider G in the upper half-plane $0\leqslant\theta\leqslant\pi$ of the physical plane. Let us first suppose that the point (r,θ) in this half-plane is in the shadow. Then $G=0$ for $t<\tau$, τ being defined by (6.2.8) and (6.2.9). The term $F(r,\theta+2m\pi,t;\,r_0)$ with $m>0$ is zero until $t=\tau+2m\pi(a/c)$. Hence it may be said to be a diffracted pulse that has encircled the cylinder m times in the positive sense. The fronts of all these pulses join up and constitute a curve that is the projection onto the physical plane of the diffracted front on $\mathscr{R}$ which propagates into $\theta>0$. The terms $F(r,\theta+2m\pi,t;\,r_0)$ with $m<0$ may similarly be interpreted as diffracted pulses that have encircled the cylinder $|m|-1$ times in the negative sense. For

$$F(r,\theta-2\pi,t;\,r_0)=F(r,2\pi-\theta,t;\,r_0)$$

arrives when

$$ct = c\tau' = (r^2 - a^2)^{\frac{1}{2}} + (r_0^2 - a^2)^{\frac{1}{2}}$$
$$+ a\left(2\pi - \theta - \cos^{-1}\frac{a}{r} - \cos^{-1}\frac{a}{r_0}\right) = c\tau + 2a(\pi - \theta),$$

which is the travel time along the path $SBQ'P'$ in fig. 6.1;

$$F(r, \theta - 2m_1\pi, t; r_0) \quad (m_1 > 1)$$

arrives when $t = \tau' + 2(m_1 - 1)\pi(a/c)$. Thus the first pulse to arrive is $F(r, \theta, t; r_0)$; it is followed by $F(r, \theta - 2\pi, t; r_0)$. The approximations that will be derived below apply only near the diffracted front, so that we shall only be able to take the first pulse into account, except near $\theta = \pi$, where the second pulse arrives with so little delay that it can also be included. At $\theta = \pi$ these two pulses arrive simultaneously and the pressure is doubled.

If the point (r, θ) is in the domain of direct reflexion then $F(r, \theta, t; r_0)$ is not a diffracted pulse; it can be shown (Friedlander, 1954, Appendix B) that it is then the sum of the incident and the reflected pulse. The other terms still represent diffracted pulses.

3. The eigenfunction expansion

The first step in the calculation of F is to take the Laplace transforms of (6.2.5) and (6.2.6). According to (5.A.16), (5.A.17) and (5.A.18) the Laplace transform $\bar{F}$ of F is

$$\bar{F}(r, \theta, s; r_0) = \int_0^\infty F(r, \theta, t; r_0)\, e^{-st}\, dt, \tag{6.3.1}$$

and (6.2.5) and (6.2.6) transform respectively into

$$\frac{\partial^2 \bar{F}}{\partial r^2} + \frac{1}{r}\frac{\partial \bar{F}}{\partial r} + \frac{1}{r^2}\frac{\partial^2 \bar{F}}{\partial \theta^2} - \frac{s^2}{c^2}\bar{F} = -\frac{1}{r}\delta(r - r_0)\,\delta(\theta), \tag{6.3.2}$$

$$\left[\frac{\partial \bar{F}}{\partial r}\right]_{r=a} = 0. \tag{6.3.3}$$

We can now determine $\bar{F}$ formally by means of the method of normal coordinates of the theory of oscillations. Let us first consider the solutions of the ordinary differential equation

$$\phi''(r) + \frac{1}{r}\phi'(r) + \left(\frac{\mu^2}{r^2} - \frac{s^2}{c^2}\right)\phi(r) = 0 \tag{6.3.4}$$

that satisfy the boundary conditions

$$\phi'(a) = 0, \quad \lim_{r\to\infty} \phi(r) = 0. \tag{6.3.5}$$

The parameter μ is a separation constant. A Laplace transform is necessarily regular in some half-plane $\mathscr{R}s > s_0$, and its values for real positive s uniquely determine its inverse. We may therefore be guided by the case of real positive s. Then (6.3.4) and (6.3.5) constitute an eigenvalue problem of a well-known and simple type with a discrete spectrum of real positive eigenvalues $\mu_1, \mu_2, \ldots$ and a complete set of eigenfunctions $\phi_1, \phi_2, \ldots$.† The solution of (6.3.4) that tends to zero as $r\to\infty$ is obviously a constant multiple of a modified Bessel function of order $i\mu$ of the second kind, $K_{i\mu}(sr/c)$. Hence the eigenvalues μ_j are the zeros of $K'_{i\mu}(sa/c)$,

$$K'_{i\mu_j}\left(\frac{sa}{c}\right) = 0 \quad (j = 1, 2, \ldots) \tag{6.3.6}$$

and the jth eigenfunction is

$$\phi_j = L_j K_{i\mu_j}\left(\frac{sr}{c}\right), \tag{6.3.7}$$

where L_j is a constant. If we denote the solution of (6.3.4) which tends to zero as $r\to\infty$ by $\phi(r, \mu)$ then

$$\frac{d}{dr}\{r[\phi'(r,\mu)\,\phi(r,\nu) - \phi(r,\mu)\,\phi'(r,\nu)]\} + \frac{\mu^2-\nu^2}{r}\,\phi(r,\mu)\,\phi(r,\nu) = 0,$$

whence

$$\int_a^\infty \phi(r,\mu)\,\phi(r,\nu)\frac{dr}{r} = \frac{a}{\mu^2-\nu^2}[\phi'(a,\mu)\,\phi(a,\nu) - \phi(a,\mu)\,\phi'(a,\nu)].$$

Hence the eigenfunctions satisfy the orthogonality relations

$$\int_a^\infty \phi_j(r)\,\phi_k(r)\frac{dr}{r} = 0 \quad (j \neq k). \tag{6.3.8}$$

By putting $\nu = \mu_j$ and making $\mu \to \mu_j$ we also deduce that

$$\int_a^\infty \phi_j^2 \frac{dr}{r} = \frac{a}{2\mu_j}\phi_j(a)\left[\frac{\partial}{\partial\mu}\phi'(a,\mu)\right]_{\mu=\mu_j}.$$

† The substitution $r = ae^z$ transforms the differential equation for ϕ into $(d^2\phi/dz^2) + [\mu^2 - (s^2/c^2)\,e^{2z}]\,\phi = 0$ and the boundary conditions into

$$[d\phi/dz]_{z=0} = 0, \quad \lim_{r\to\infty} \phi = 0.$$

For the theory of such eigenvalue problems see for instance Titchmarsh (1946), pp. 113–16.

Hence the eigenfunctions

$$\phi_j(r)=\left\{\frac{c}{sa}\frac{2\mu_j}{K_{1\mu_j}\left(\frac{sa}{c}\right)\left[\frac{\partial}{\partial\mu}K'_{1\mu}\left(\frac{sa}{c}\right)\right]_{\mu=\mu_j}}\right\}^{\frac{1}{2}}K_{1\mu_j}\left(\frac{sr}{c}\right) \qquad (6.3.9)$$

satisfy (6.3.8) and the normalization condition

$$\int_a^\infty \phi_j^2\frac{dr}{r}=1. \qquad (6.3.10)$$

Any suitable function $f(r)$ that satisfies $f'(a)=0$ can be expanded as an eigenfunction series

$$f(r_0)\sim\sum_{j=1}^{\infty}\phi_j(r_0)\int_a^\infty f(r)\,\phi_j(r)\frac{dr}{r}.$$

This implies formally that

$$\frac{1}{r}\sum_{j=1}^{\infty}\phi_j(r_0)\,\phi_j(r)=\delta(r-r_0). \qquad (6.3.11)$$

We may therefore solve (6.3.2) by putting

$$\bar{F}=\sum_{j=1}^{\infty}\bar{F}_j$$

with $$\frac{\partial^2\bar{F}_j}{\partial r^2}+\frac{1}{r}\frac{\partial\bar{F}_j}{\partial r}+\frac{1}{r^2}\frac{\partial^2\bar{F}_j}{\partial\theta^2}-\frac{s^2}{c^2}\bar{F}_j=-\frac{1}{r^2}\phi_j(r_0)\,\phi_j(r)\,\delta(\theta).$$

If we assume that $\bar{F}_j=\phi_j(r_0)\,\phi_j(r)\,\chi_j(\theta)$, then

$$\chi_j''-\mu_j^2\chi_j=-\delta(\theta),$$

since ϕ_j satisfies (6.3.4) with $\mu=\mu_j$. Thus χ_j is the Green's function of the operator $\mu_j^2-(d^2/d\theta^2)$. On $\theta\neq 0$, χ_j satisfies the homogeneous equation $\chi_j''-\mu_j^2\chi_j=0$ and at $\theta=0$

$$\chi_j(+0)=\chi_j(-0), \quad \chi_j'(+0)-\chi_j'(-0)=-1.$$

Hence $\chi_j=(1/2\mu_j)\exp(-\mu_j|\theta|)$ if we add the condition that $\chi_j\to 0$ as $|\theta|\to\infty$. We therefore have the following eigenfunction expansion of $\bar{F}$:

$$\left.\begin{aligned}\bar{F}(r,\theta,s;r_0)&=\sum_{j=1}^{\infty}\bar{F}_j(r,\theta,s;r_0),\\ \bar{F}_j=\frac{\phi_j(r_0)\,\phi_j(r)}{2\mu_j}e^{-\mu_j|\theta|}&=\frac{c}{sa}\frac{K_{1\mu_j}\left(\frac{sr_0}{c}\right)K_{1\mu_j}\left(\frac{sr}{c}\right)}{K_{1\mu_j}\left(\frac{sa}{c}\right)\left[\frac{\partial}{\partial\mu}K'_{1\mu}\left(\frac{sa}{c}\right)\right]_{\mu=\mu_j}}e^{-\mu_j|\theta|}.\end{aligned}\right\} \qquad (6.3.12)$$

This formal procedure can be justified for real positive s.† It is shown in the appendix to this chapter that the series (6.3.12) converges also for complex s in $\mathscr{R}s > 0$ provided that $\theta \neq 0$. It may therefore be reasonably assumed to be the Laplace transform of the Green's function F.

The functions $\phi_j(r)\exp(\pm\mu_j\theta)$ are obviously solutions of the Laplace-transformed wave equation that satisfy the boundary condition at the cylinder and are single-valued on the Riemann surface $\mathscr{R}$; they are the eigenfunctions of the Laplace-transformed problem. In the theory of radio waves (where $s = i\omega$, ω being the frequency) such eigenfunctions are called *propagation modes*. It will be shown in the next section that in the shadow each $\bar{F}_j$ is the Laplace transform of a pulse which may be called a pulse propagation mode. In the case of radio waves, expansions in propagation modes are found to be suitable for the calculation of diffracted fields. This is also true in the pulse case. In the domain of direct reflexion the series (6.3.12) converges but does not lend itself to the calculation of the reflected field. For this purpose an integral representation of $\bar{F}$ must be used which can be deduced from (6.3.12) by a technique applicable to all problems of this type.‡ But in the present case it can be obtained more simply by taking the Fourier transform of (6.3.2) with respect to θ. The details will be found in Friedlander (1954).

4. The diffraction formulae

It is obviously not possible to evaluate the inverse Laplace transform of (6.3.12) as it stands. But we are primarily concerned with the behaviour of the diffracted pulse near its front. It is well known that this is correlated with the behaviour of its Laplace transform for large s. We must therefore examine the series in this case. In order to obtain asymptotic formulae for the individual modes one has to derive approximate solutions of the differential

† $\bar{F}$ will be a solution of (6.3.2) in the sense of the theory of distributions if both the series (6.3.11) and (6.3.12) converge in the weak sense. This means that the series obtained on multiplication by a test-function $f(r)$ and integration from a to ∞ converge and that the sum of the first series is then $f(r_0)$. The proof of these statements is an immediate consequence of the properties of the eigenfunctions; see Titchmarsh (1946), p. 26, Theorem 2.7.

‡ Marcuvitz (1951). A detailed exposition of the subject is given in Friedman (1956), ch. 5.

equation (6.3.4) valid when μ and s are large. From these the manner in which the μ_j depend on s as $s \to \infty$ can be deduced, and asymptotic formulae for the functions in the numerator and denominator of $\bar{F}_j$ can then be derived. This programme can be carried out by means of standard methods of approximation, but as it involves a considerable amount of work which is irrelevant to the main argument the calculation will be omitted; it is given in the appendix to this chapter. It is shown there that

$$\bar{F}_j(r, \theta, s; r_0) \sim \frac{a^{\frac{1}{3}} c^{\frac{2}{3}}}{2^{\frac{7}{3}} \pi \alpha_j [\mathrm{Ai}(-\alpha_j)]^2} \frac{s^{-\frac{2}{3}}}{(r^2 - a^2)^{\frac{1}{4}} (r_0^2 - a^2)^{\frac{1}{4}}} \times \exp\left\{-s\tau - \left(\frac{a}{2c}\right)^{\frac{1}{3}} \alpha_j \delta s^{\frac{1}{3}}\right\}, \quad (6.4.1)$$

where $-\alpha_j$ denotes the jth zero of $\mathrm{Ai}'(\alpha)$, $\mathrm{Ai}(\alpha)$ being the Airy function, and τ, δ are defined by (6.2.8) and (6.2.9). This approximation breaks down if $r = a$, $r_0 = a$ or $r = r_0 = a$. In these cases the factor multiplying the exponential must be modified:

$$\bar{F}_j(a, \theta, s; r_0) \sim \frac{c^{\frac{1}{2}}}{2^{\frac{3}{2}} \pi \alpha_j \, \mathrm{Ai}(-\alpha_j)} \frac{s^{-\frac{1}{2}}}{(r_0^2 - s^2)^{\frac{1}{4}}} \exp\left(-s\tau - \left(\frac{a}{2c}\right)^{\frac{1}{3}} \alpha_j \delta s^{\frac{1}{3}}\right), \quad (6.4.2)$$

$$\bar{F}_j(\alpha, \theta, s; a) \sim \frac{1}{2^{\frac{2}{3}} \alpha_j} \left(\frac{c}{sa}\right)^{\frac{1}{3}} \exp\left(-s\tau - \left(\frac{a}{2c}\right)^{\frac{1}{3}} \alpha_j \delta s^{\frac{1}{3}}\right). \quad (6.4.3)$$

Also $\bar{F}_j(r, \theta, s; a) = \bar{F}_j(a, \theta, s; r)$, so that this case is covered by (6.4.2). The reason why different approximations must be employed is that the cylinder is a caustic of the diffracted rays; the factor $(r^2 - a^2)^{-\frac{1}{4}}$ is a divergence factor which satisfies the transport equation associated with the diffracted front.†

Each mode $\bar{F}_j$ contains the 'delay factor' $e^{-s\tau}$ which implies that the inverse Laplace transform (if it exists) is zero for $t < \tau$. Since τ is the arrival time of the diffracted front this suggests at once that the series can be used to calculate the diffracted pulse. $\bar{F}_j$ is obviously a Laplace transform in the shadow where $\delta > 0$ but not in the domain of direct reflexion where $\delta < 0$; for in the latter

† The complete asymptotic expansion is obtained by multiplying (6.4.1)–(6.4.3) by series in descending powers of $s^{\frac{1}{3}}$. Such expansions are investigated from a formal point of view in a paper by J. B. Keller and the author (Keller and Friedlander, 1955). It is shown there that the coefficient of the leading term satisfies the transport equation.

case $e^{s\tau}\bar{F}_j$ is not regular at infinity in $\mathscr{R}s>0$. It seems therefore reasonable to assume that the Laplace inversion of our series can be carried out term by term in the shadow.

Now

$$\bar{F}_j \sim s^{-k} \exp\left\{-s\tau - \left(\frac{a}{2c}\right)^{\frac{1}{3}} \alpha_j \delta s^{\frac{1}{3}}\right\} \sum_{m=0}^{\infty} L_m^{(j)} s^{-\frac{1}{3}m},$$

where k is either $\frac{2}{3}$, $\frac{1}{2}$ or $\frac{1}{3}$, and the $L_m^{(j)}$ depend on r and θ; we know of course only the leading term $L_0^{(j)}$. Hence

$$F_j \sim \sum_{m=0}^{\infty} L_m^{(j)} U_{k+\frac{1}{3}m}\left\{\left(\frac{a}{2c}\right)^{\frac{1}{3}} \alpha_j \delta,\, T\right\}, \tag{6.4.4}$$

where we have put

$$T = t - \tau \tag{6.4.5}$$

for the time counted from the arrival of the diffracted front, and

$$U_n(\beta, T) = \frac{1}{2\pi i}\int_{G-i\infty}^{G+i\infty} s^{-n} \exp(Ts - \beta s^{\frac{1}{3}})\, ds, \tag{6.4.6}$$

G being a positive constant. If $3n$ is an integer U_n can be expressed in terms of the Airy function. But since we are already dealing with approximations and also have to consider cases in which $3n$ is not an integer it is best to adopt a uniform approach and approximate U_n for small T by the method of steepest descents.

The function $Ts - \beta s^{\frac{1}{3}}$ has a saddle point

$$s = s_0 = \left(\frac{\beta}{3T}\right)^{\frac{3}{2}}$$

on the positive real axis. It can be shown that the contour can be deformed into the path of steepest descent through this saddle point. On this path we put

$$Ts - \beta s^{\frac{1}{3}} = -\frac{2}{3^{\frac{3}{2}}}\frac{\beta^{\frac{3}{2}}}{T^{\frac{1}{2}}} - \zeta^2,$$

where ζ is real. Dividing by $s_0 T$ we have

$$\frac{s}{s_0} - 3\left(\frac{s}{s_0}\right)^{\frac{1}{3}} + 2 = -\frac{2\zeta^2}{\xi}, \quad \xi = \frac{2}{3^{\frac{3}{2}}}\frac{\beta^{\frac{3}{2}}}{T^{\frac{1}{2}}},$$

whence

$$\left(\frac{s}{s_0} - 1\right)\left\{1 - \frac{5}{9}\left(\frac{s}{s_0} - 1\right) + \frac{10}{27}\left(\frac{s}{s_0} - 1\right)^2 - \dots\right\}^{\frac{1}{2}} = i\left(\frac{6}{\xi}\right)^{\frac{1}{2}} \zeta.$$

Extracting the square root and reversing the series one finds

$$s = s_0\left\{1 + \mathrm{i}\left(\frac{6}{\xi}\right)^{\frac{1}{2}}\zeta - \frac{5}{3\xi}\zeta^2 - \frac{5\mathrm{i}}{108}\frac{6^{\frac{1}{2}}}{\xi^{\frac{3}{2}}}\zeta^3 + \dots\right\},$$

$$U_n \sim \frac{s_0^{1-n}}{2\pi}\left(\frac{6}{\xi}\right)^{\frac{1}{2}} \mathrm{e}^{-\xi}\int_{-\infty}^{\infty} \mathrm{e}^{-\zeta^2}\left\{1 + \frac{(5-9n)\,\mathrm{i}}{9}\left(\frac{6}{\xi}\right)^{\frac{1}{2}}\zeta - \frac{108n^2 - 72n + 5}{36\xi}\zeta^2 + \dots\right\}\mathrm{d}\zeta,$$

whence

$$U_n \sim \frac{3^{\frac{1}{4}(6n-1)}}{2\pi^{\frac{1}{2}}}\beta^{\frac{1}{4}(3-6n)}T^{\frac{1}{4}(6n-5)}H(T)\exp\left\{-\frac{2\beta^{\frac{3}{2}}}{3^{\frac{3}{2}}T^{\frac{1}{2}}}\right\} \times\left\{1 - \frac{108n^2 - 72n + 5}{72\xi} + \dots\right\}, \quad (6.4.7)$$

where a factor $H(T)$ has been inserted because $U_n = 0$ for $T < 0$.†

We must substitute this in (6.4.4). Since only the leading coefficient $L_0^{(j)}$ is known the resulting approximation is

$$F_j = \delta^{\frac{1}{4}(3-6k)}T^{\frac{1}{4}(6k-5)}H(T)\exp\left\{-\left(\frac{\alpha_j\delta}{3}\right)^{\frac{3}{2}}\left(\frac{2a}{cT}\right)^{\frac{1}{2}}\right\} \times\left\{K_0^{(j)}\left[1 + O\left(\frac{T^{\frac{1}{2}}}{\delta^{\frac{3}{2}}}\right)\right] + O\left(\frac{T^{\frac{1}{2}}}{\delta^{\frac{1}{2}}}\right)\right\},$$

where
$$K_0^{(j)} = \frac{3^{\frac{1}{4}(6k-1)}}{2\pi^{\frac{1}{2}}}\left(\frac{a}{2c}\right)^{\frac{1}{4}(1-k)}\alpha_j^{\frac{1}{4}(3-6k)}L_0^{(j)}.$$

The magnitude of the error term $O(T^{\frac{1}{2}}\delta^{-\frac{3}{2}})$ (which predominates when the shadow boundary is approached) can be inferred from (6.4.7), but the constant implied by the $O(T^{\frac{1}{2}}\delta^{-\frac{1}{2}})$ is unknown. The errors will of course be neglected in the final formula. This has an important consequence. Since $\alpha_1 < \alpha_2 < \dots$, the leading term of any mode of order $j > 1$ tends to zero more rapidly than the error term of the first mode. Since this is to be neglected it follows that we must confine ourselves to a range of T in which the higher modes are negligible and retain only the first mode. This is an additional reason why our approximations cannot be used as the shadow boundary is approached ($\delta \to 0$). Our final approximations,

† For $T < 0$ the contour can be deformed into a large semicircle in $\mathscr{R}s > 0$ the integral along which vanishes in the limit by Jordan's lemma.

obtained by combining (6.4.7) with (6.4.1)–(6.4.3), are therefore as follows:

$$F(r,\theta,t;r_0)\sim A\frac{c}{a}\left\{\frac{a^4}{(r^2-a^2)(r_0^2-a^2)}\right\}^{\frac{1}{4}}$$

$$\times\left(\frac{a}{cT\delta}\right)^{\frac{1}{4}}H(T)\,\mathrm{e}^{-\xi}\quad (r>a,\ r_0>a),\qquad (6.4.8)$$

$$F(a,\theta,t;r_0)\sim B\frac{c}{a}\left(\frac{a^2}{r_0^2-a^2}\right)^{\frac{1}{4}}\left(\frac{a}{cT}\right)^{\frac{1}{2}}H(T)\,\mathrm{e}^{-\xi},\qquad (6.4.9)$$

$$F(a,\theta,t;a)\sim C\frac{c}{a}\delta^{\frac{1}{4}}\left(\frac{a}{cT}\right)^{\frac{3}{4}}H(T)\,\mathrm{e}^{-\xi}.\qquad (6.4.10)$$

Here
$$\xi=\frac{2^{\frac{1}{2}}\alpha_1^{\frac{3}{2}}}{3^{\frac{3}{2}}}\delta^{\frac{3}{2}}\left(\frac{a}{cT}\right)^{\frac{1}{2}}=0{\cdot}280\,\delta^{\frac{3}{2}}\left(\frac{a}{cT}\right)^{\frac{1}{2}},\qquad (6.4.11)$$

and A, B, C are constants,

$$\left.\begin{aligned}A&=\frac{6^{\frac{3}{4}}}{16\pi^{\frac{3}{2}}\alpha_1^{\frac{5}{4}}[\mathrm{Ai}(-\alpha_1)]^2}=0{\cdot}147,\\ B&=\frac{6^{\frac{1}{2}}}{8\pi\alpha_1\,\mathrm{Ai}(-\alpha_1)}=0{\cdot}179,\quad C=\frac{6^{\frac{1}{4}}}{4\pi^{\frac{1}{2}}\alpha_1^{\frac{3}{4}}}=0{\cdot}218.\dagger\end{aligned}\right\}\qquad (6.4.12)$$

The time taken by a wave front to travel a distance equal to the radius of the cylinder, a/c, is a characteristic time of our problem. By (6.2.7) the dimension of an elementary solution is the reciprocal of time. Hence Fa/c is a function of r/a, r_0/a and cT/a; the formulae (6.4.8)–(6.4.10) have been arranged accordingly. They are all diffraction formulae of the type (6.1.1).

It remains to substitute the approximations in the expansion (6.2.4) of the Green's function G of the cylinder in the physical plane. The terms of this series represent diffracted pulses that have encircled the cylinder a number of times. Since the diffraction formulae are valid for small T, we can only take the first pulse that arrives into account, except near $\theta=\pi$, where the second pulse may be included. Taking $\theta>0$ the first pulse is $F(r,\theta,t;r_0)$ and the second one $F(r,\theta-2\pi,t;r_0)$; their arrival times are respectively

$$c\tau=(r^2-a^2)^{\frac{1}{2}}+(r_0^2-a^2)^{\frac{1}{2}}+a\delta,\quad \delta=\theta-\cos^{-1}\frac{a}{r}-\cos^{-1}\frac{a}{r_0},\qquad (6.4.13)$$

$$c\tau'=(r^2-a^2)^{\frac{1}{2}}+(r_0^2-a^2)^{\frac{1}{2}}+a\delta',\quad \delta'=2\pi-\theta-\cos^{-1}\frac{a}{r}-\cos^{-1}\frac{a}{r_0},\qquad (6.4.14)$$

† $\alpha_1=1{\cdot}0188$, $\mathrm{Ai}(-\alpha_1)=0{\cdot}5357$ (J. C. P. Miller, 1946).

so that the second pulse arrives with the time delay $(2a/c)(\pi-\theta)$ relative to the first one. We therefore write

$$G(r,\theta,t;\,r_0)=F(r,\theta,t;\,r_0)+F(r,\theta-2\pi,t;\,r_0), \qquad (6.4.15)$$

with the understanding that $F(r,\theta-2\pi,t;\,r_0)$ is to be ignored unless $\pi-\theta$ is sufficiently small.

5. Diffraction of a plane pulse

The diffraction formulae for an incident plane pulse can be deduced from those for the Green's function. According to the superposition principle it is sufficient to consider the special case of an incident 'unit pressure pulse'

$$p_1=H(ct+r\cos\theta). \qquad (6.5.1)$$

The Laplace transform of this is

$$\bar{p}_1=\frac{1}{s}\exp\left(\frac{sr}{c}\cos\theta\right). \qquad (6.5.2)$$

Now the Laplace transform of the elementary solution in free space, (6.2.7), is

$$\bar{E}=\frac{1}{2\pi}K_0\left(\frac{sR}{c}\right)=\frac{1}{2\pi}\left(\frac{\pi c}{2sR}\right)^{\frac{1}{2}}e^{-sR/c}\left\{1+O\left(\frac{1}{sR}\right)\right\},$$

where $$R=(r^2+r_0^2-2rr_0\cos\theta)^{\frac{1}{2}}.$$

When r_0 is large then $R=r_0-r\cos\theta+O(1/r_0)$, so that

$$\bar{E}=\frac{1}{2\pi}\left(\frac{\pi c}{2sr_0}\right)^{\frac{1}{2}}e^{-sr_0/c}\exp\left(\frac{sr}{c}\cos\theta\right)\left\{1+O\left(\frac{1}{r_0}\right)\right\}.$$

Hence $$\lim_{r_0\to\infty}\left\{2\pi\left(\frac{2r_0}{\pi cs}\right)^{\frac{1}{2}}e^{sr_0/c}\,\bar{E}\right\}=\bar{p}_1. \qquad (6.5.3)$$

The Laplace transform of the diffracted field $P(r,\theta,t)$ due to the incident pulse (6.5.1) is therefore

$$\bar{P}(r,\theta,s)=\lim_{r_0\to\infty}\left\{2\pi\left(\frac{2r_0}{\pi cs}\right)^{\frac{1}{2}}e^{sr_0/c}\,G(r,\theta,s;\,r_0)\right\}. \qquad (6.5.4)$$

By (6.2.4) $$P(r,\theta,t)=\sum_{m=-\infty}^{\infty}Q(r,\theta+2m\pi,t;\,r_0), \qquad (6.5.5)$$

where Q is the inverse Laplace transform of

$$\bar{Q}(r,\theta,s) = \lim_{r_0\to\infty}\left\{2\pi\left(\frac{2r_0}{\pi cs}\right)^{\frac{1}{2}} e^{sr_0/c}\bar{F}(r,\theta,s;\,r_0)\right\}. \qquad (6.5.6)$$

Substituting the eigenfunction expansion (6.3.12) for $\bar{F}$ we obtain

$$\bar{Q}(r,\theta,s) = \frac{2\pi c}{s^2 a}\sum_{j=1}^{\infty}\frac{K_{1\mu_j}(sr/c)}{K_{1\mu_j}(sa/c)\left[\frac{\partial}{\partial\mu}K_{1\mu}(sa/c)\right]_{\mu=\mu_j}} e^{-\mu_j|\theta|}. \qquad (6.5.7)$$

This series converges for $|\theta| > \frac{1}{2}\pi$ in the half-plane $\mathscr{R}s > 0$. The asymptotic formulae for Q can be deduced by means of the approximations derived in the appendix, but they can also be obtained directly by applying the limiting procedure (6.5.6) to (6.4.1) and (6.4.2). We note first that

$$\lim_{r_0\to\infty}\left(\tau - \frac{r_0}{c}\right) = \sigma = \frac{1}{c}(r^2-a^2)^{\frac{1}{2}} + \frac{a}{c}\gamma, \qquad (6.5.8)$$

$$\lim_{r_0\to\infty}\delta \qquad = \gamma = |\theta| - \tfrac{1}{2}\pi - \cos^{-1}\frac{a}{r}. \qquad (6.5.9)$$

The glancing incident rays are $y = \pm a$. They touch the cylinder at $(a, \frac{1}{2}\pi)$ and $(a, -\frac{1}{2}\pi)$, so that the shadow on the Riemann surface contains the two arcs of the circle $r=a$ on which $|\theta| > \frac{1}{2}\pi$. Hence γ has a meaning similar to that of δ; it is the angle subtended at the origin by the foot of the diffracted ray and the appropriate point of the shadow boundary on the cylinder. This implies that $\gamma > 0$ in the shadow. The quantity σ is the arrival time of the diffracted front. When the formulae (6.4.1) and (6.4.2) with $j=1$ are substituted for $\bar{F}$ in (6.5.6) we find that

$$\bar{Q}(r,\theta,s) \sim \frac{2^{-\frac{5}{6}}c^{\frac{1}{6}}a^{\frac{1}{3}}}{\pi^{\frac{1}{2}}\alpha_1[\mathrm{Ai}(-\alpha_1)]^2}\frac{s^{-\frac{7}{6}}}{(r^2-a^2)^{\frac{1}{4}}}\exp\left(-s\sigma - \left(\frac{a}{2c}\right)^{\frac{1}{3}}\alpha_1\gamma s^{\frac{1}{3}}\right) \qquad (6.5.10)$$

for $r > a$, and

$$\bar{Q}(a,\theta,s) \sim \frac{1}{\alpha_1\,\mathrm{Ai}(-\alpha_1)\,s}\exp\left(-\sigma s - \left(\frac{a}{2c}\right)^{\frac{1}{3}}\alpha_1\gamma s^{\frac{1}{3}}\right), \qquad (6.5.11)$$

since as in the case of the Green's function only the first mode can be taken into account. The diffraction formulae valid for small T, now defined by

$$T = t - \sigma, \tag{6.5.12}$$

are then found by means of (6.4.7):

$$Q(r,\theta,t) \sim A' \frac{a^{\frac{1}{2}}}{(r^2-a^2)^{\frac{1}{4}}} \frac{1}{\gamma} \left(\frac{cT}{a}\right)^{\frac{1}{2}} H(T)\,\mathrm{e}^{-\xi}, \tag{6.5.13}$$

$$Q(a,\theta,t) \sim B' \gamma^{-\frac{3}{4}} \left(\frac{cT}{a}\right)^{\frac{1}{4}} H(T)\,\mathrm{e}^{-\xi}, \tag{6.5.14}$$

where ξ is given by (6.4.11) with δ replaced by γ, and

$$A' = \frac{3^{\frac{3}{2}}}{2^{\frac{3}{2}}\pi[\alpha_1 \operatorname{Ai}(-\alpha_1)]^2} = 1{\cdot}926, \quad B' = \frac{3^{\frac{5}{4}}}{2^{\frac{3}{4}}\pi^{\frac{1}{2}}\alpha_1^{\frac{7}{4}} \operatorname{Ai}(-\alpha_1)} = 2{\cdot}347. \tag{6.5.15}$$

Finally we have

$$P(r,\theta,t) = Q(r,\theta,t) + Q(r,\theta-2\pi,t), \tag{6.5.16}$$

where, for $\theta > 0$, $Q(r,\theta-2\pi,t)$ is to be ignored unless $\pi-\theta$ is sufficiently small.

The Laplace transform of the incident pulse

$$f\left(t + \frac{r}{c}\cos\theta\right) \tag{6.5.17}$$

is (6.5.2) multiplied by $s\bar{f}(s)$, where $\bar{f}(s)$ is the Laplace transform of $f(t)$. If we suppose that $f(t) \sim \frac{t^k}{k!} H(t)$ when t is small, then $\bar{f}(s) \sim s^{-k-1}$. By (6.4.7), multiplication of (6.5.10) and (6.5.11) by s^{-k} is equivalent to multiplication of (6.5.13) and (6.5.14) by the factor

$$\left(\frac{2c}{a}\right)^{\frac{1}{3}k} \left(\frac{3T}{\alpha_1\gamma}\right)^{\frac{2}{3}k}. \tag{6.5.18}$$

Some pressure-time curves on the cylinder are shown in fig. 6.2, where P has been plotted against cT/a for various values of γ from (6.5.14). The right-hand side of (6.5.14) is a function of ξ which is proportional to $T^{\frac{1}{3}}\gamma^{-\frac{3}{2}}$ only, so that the curve at γ' is obtained from that at γ by changing the time-scale in the ratio $(\gamma'/\gamma)^3$. Hence the pressure-time curves steepen as γ decreases. The curve for $\gamma = 90°$ is not shown; because of the doubling of

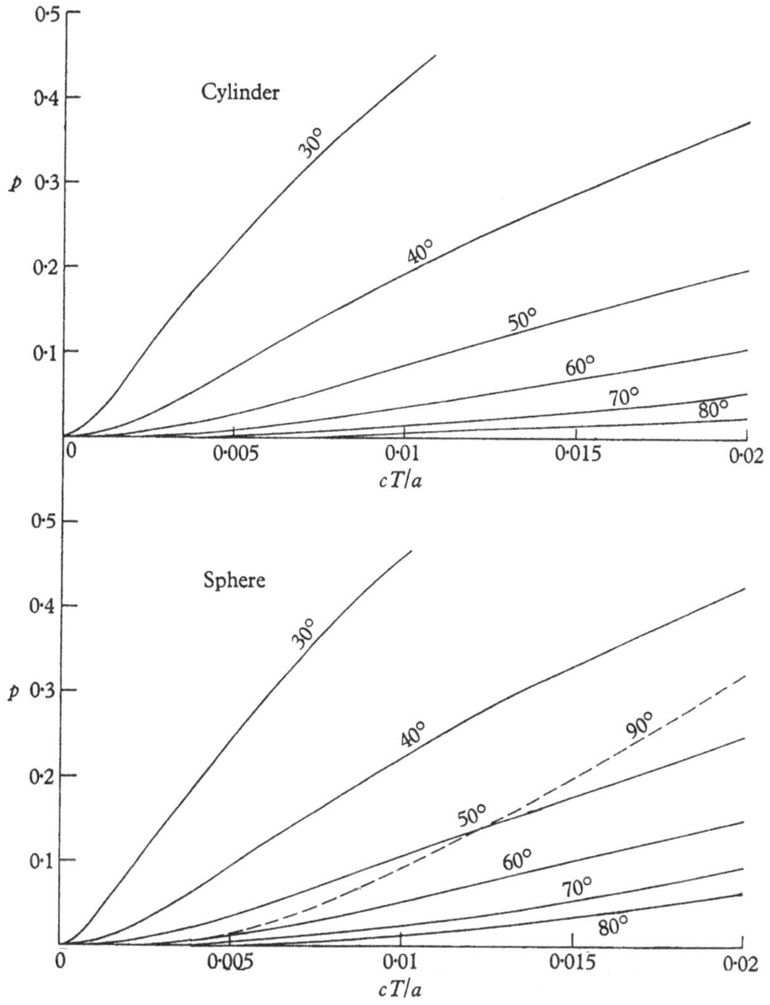

Fig. 6.2. Curves of pressure against time T counted from the onset of the diffracted pulse for various values of γ, (*a*) on the surface of a circular cylinder and (*b*) on a sphere. The incident pulse is a plane unit pressure pulse.

pressure at that point due to the simultaneous arrival of the two pressure pulses on the right-hand side of (6.5.16) it is undistinguishable from the curve for $\gamma = 80°$. The same information is displayed in a different manner in fig. 6.3, where the pressure P is plotted against γ for constant cT/a.

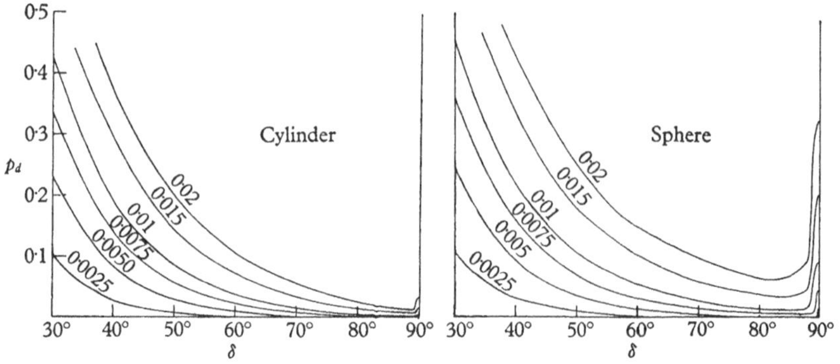

Fig. 6.3. The curves of fig. 6.2 re-plotted as curves of pressure against γ for constant T, (a) on the circular cylinder and (b) on the sphere.

6. The Green's function of the sphere

As a second example of pulse diffraction by a continuously curved obstacle we consider the case of a fixed and rigid sphere. This can be solved by the same technique as the problem of the circular cylinder. Taking the centre O of the sphere as origin, let r now denote the distance from O and θ, ϕ the co-latitude and longitude respectively. The radius of the sphere will be denoted by a. By symmetry it is sufficient to consider a source at $r=r_0$ and $\theta=0$, with $r_0 \geqslant a$. The geometrical optics of the problem is similar to that of the cylinder; in fact, the curves of fig. 6.1 are now the meridional sections of the fronts, and the actual fronts are obtained by rotating them about their axis of symmetry. Hence there is one important new feature: *the portion $\theta=\pi$ of the axis in the shadow is now a caustic of the diffracted fronts.* Consequently, different representations of the field are required for $\theta<\pi$ and for $\theta=\pi$.

We cannot use the Riemann-surface device in this problem but must confine θ to the interval $0 \leqslant \theta \leqslant \pi$. It is not difficult to obtain solutions of problems involving spherical boundaries as series expansions in surface harmonics of integral order. These correspond to Fourier series in the cylindrical case and, like them, are not suitable for the discussion of the diffracted field. The appropriate representation of the Laplace transform of the Green's function can be obtained from this series by a transformation due to Watson (1919). But we shall use the method of eigenfunctions instead. To avoid certain minor analytical difficulties we shall first assume that we are dealing with a 'ring source' on $r=r_0$, $\theta=\theta_0$ and then make $\theta_0 \to 0$.

In order to obtain the delta-function representation δ^* of the ring source we note that

$$\int_0^\infty \int_0^\pi f(r,\theta)\, \delta^*(r,\theta;\, r_0,\theta_0)\, 2\pi r^2 \sin\theta \,\mathrm{d}r\,\mathrm{d}\theta = f(r_0,\theta_0)$$

for any test-function f with axial symmetry. Hence we can put

$$\delta^* = \frac{\delta(r-r_0)\,\delta(\theta-\theta_0)}{2\pi r^2 \sin\theta},$$

and formulate the wave equation for the Green's function G of our problem as

$$\frac{1}{c^2}\frac{\partial^2 G}{\partial t^2} - \left\{\frac{\partial^2 G}{\partial r^2} + \frac{2}{r}\frac{\partial G}{\partial r} + \frac{1}{r^2 \sin\theta}\frac{\partial}{\partial\theta}\left(\sin\theta \frac{\partial G}{\partial\theta}\right)\right\} = \frac{\delta(r-r_0)\,\delta(\theta-\theta_0)\,\delta(t)}{2\pi r^2 \sin\theta}. \quad (6.6.1)$$

G is now a distribution compact towards the past which is the derivative of a function with respect to t. We need not discuss the formulation of the boundary condition at the sphere, since we shall determine G as the inverse of its Laplace transform $\bar{G}$. This satisfies

$$\frac{\partial^2 \bar{G}}{\partial r^2} + \frac{2}{r}\frac{\partial \bar{G}}{\partial r} + \frac{1}{r^2 \sin\theta}\frac{\partial}{\partial\theta}\left(\sin\theta \frac{\partial \bar{G}}{\partial\theta}\right) - \frac{s^2}{c^2}\bar{G} = -\frac{\delta(r-r_0)\,\delta(\theta-\theta_0)}{2\pi r^2 \sin\theta}, \quad (6.6.2)$$

and the boundary condition

$$\left[\frac{\partial \bar{G}}{\partial r}\right]_{r=a} = 0. \quad (6.6.3)$$

The first step in the calculation of $\bar{G}$ is the determination of the eigenfunctions of the problem that satisfy the homogeneous version of (6.6.2) and the boundary condition. The method of separation of variables leads at once to the result that they are of the form $\psi(r,\mu)\,\chi(\theta,\mu)$, where μ is a separation constant and

$$\frac{\mathrm{d}^2\psi}{\mathrm{d}r^2} + \frac{2}{r}\frac{\mathrm{d}\psi}{\mathrm{d}r} + \left(\frac{\mu^2+\frac{1}{4}}{r^2} - \frac{s^2}{c^2}\right)\psi = 0, \quad \left[\frac{\mathrm{d}\psi}{\mathrm{d}r}\right]_{r=a} = 0, \quad (6.6.4)$$

$$\frac{1}{\sin\theta}\frac{\mathrm{d}}{\mathrm{d}\theta}\left(\sin\theta \frac{\mathrm{d}\chi}{\mathrm{d}\theta}\right) - (\mu^2+\tfrac{1}{4})\,\chi = 0. \quad (6.6.5)$$

The separation constant μ is the parameter whose eigenvalues must be determined from (6.6.4) supplemented by the 'radiation condition' $\lim_{r\to\infty}\psi=0$. The solution of the differential equation (6.6.4) that satisfies this is

$$\psi = Lr^{-\frac{1}{2}}K_{1\mu}\left(\frac{sr}{c}\right), \tag{6.6.6}$$

where L is a constant. The eigenvalues μ_j are then determined by the boundary condition at $r=a$,

$$K'_{1\mu_j}\left(\frac{sa}{c}\right) - \frac{c}{2sa}K_{1\mu_j}\left(\frac{sa}{c}\right) = 0. \tag{6.6.7}$$

This can be treated in the same way as the condition (6.3.6) in the cylinder problem. By (6.A.23) and (6.A.24) $(c/2sa)\,K_{1\mu_j}(sa/c)$ is negligible in comparison with $K'_{1\mu_j}(sa/c)$ in the critical range

$$\mu - \frac{sa}{c} \sim \left(\frac{sa}{c}\right)^{\frac{1}{3}}$$

when s is large. Hence for large s and fixed j the μ_j are again given by (6.A.26),

$$\mu_j = \frac{sa}{c} + \alpha_j\left(\frac{sa}{2c}\right)^{\frac{1}{3}} + O(s^{-\frac{1}{3}}).$$

We shall therefore be able to use the approximations derived in §4 of the appendix again.

The orthogonality relations satisfied by the eigenfunctions ψ_j can be deduced in the usual way from (6.6.4). If $\psi(r,\mu)$ and $\psi(r,\nu)$ are solutions of (6.6.4) and of the corresponding equation with ν instead of μ respectively that both tend to zero as $r\to\infty$ then

$$\frac{d}{dr}\{r^2[\psi'(r,\mu)\,\psi(r,\nu) - \psi(r,\mu)\,\psi'(r,\nu)]\} + (\mu^2-\nu^2)\,\psi(r,\mu)\,\psi(r,\nu) = 0,$$

whence

$$\int_a^\infty \psi(r,\mu)\,\psi(r,\nu)\,dr = \frac{a^2}{\mu^2-\nu^2}[\psi'(a,\mu)\,\psi(a,\nu) - \psi(a,\mu)\,\psi'(a,\nu)]. \tag{6.6.8}$$

Putting $\mu=\mu_j$, $\nu=\mu_k$, $\psi(r,\mu_j)=\psi_j(r)$, $\psi(r,\mu_k)=\psi_k(r)$, we find

$$\int_a^\infty \psi_j\psi_k\,dr = 0 \quad (j \neq k). \tag{6.6.9}$$

The proper normalization condition is now

$$\int_a^\infty \psi_j^2\,dr = 1 \quad (j=1,2,\ldots), \tag{6.6.10}$$

and the formal completeness relation (for test-functions) is

$$\delta(r-r_0)=\sum_{j=1}^{\infty}\psi_j(r_0)\,\psi_j(r). \tag{6.6.11}$$

In order to determine the normalized eigenfunctions we put $\nu=\mu_j$ in (6.6.8) and make $\mu\to\mu_j$. Then

$$1=\int_a^{\infty}\psi_j^2\,\mathrm{d}r=\frac{a^2}{2\mu_j}\psi_j(a)\left[\frac{\partial}{\partial\mu}\psi'(a,\mu)\right]_{\mu=\mu_j}.$$

Hence
$$\psi_j=L_j r^{-\frac{1}{2}}K_{1\mu_j}\left(\frac{sr}{c}\right), \tag{6.6.12}$$
where
$$\frac{1}{L_j^2}=\frac{sa}{2c\mu_j}K_{1\mu_j}\left(\frac{sa}{c}\right)\left[\frac{\partial}{\partial\mu}K'_{1\mu}\left(\frac{sa}{c}\right)-\frac{c}{sa}K_{1\mu}\left(\frac{sa}{c}\right)\right]_{\mu=\mu_j}. \tag{6.6.13}$$

We now substitute

$$\bar{G}=\sum_{j=1}^{\infty}\psi_j(r_0)\,\psi_j(r)\,\chi_j(\theta,\theta_0) \tag{6.6.14}$$

in the differential equation (6.6.2) and replace $\delta(r-r_0)$ in the right-hand side of that equation by its expansion (6.6.11). Since ψ_j satisfies (6.6.4) with $\mu=\mu_j$ this gives

$$\sum_{j=1}^{\infty}\psi_j(r_0)\,\psi_j(r)\left\{\frac{1}{\sin\theta}\frac{\mathrm{d}}{\mathrm{d}\theta}\left(\sin\theta\frac{\mathrm{d}\chi_j}{\mathrm{d}\theta}\right)-(\mu_j^2+\tfrac{1}{4})\,\chi_j\right\}$$
$$=-\frac{\delta(\theta-\theta_0)}{2\pi\sin\theta}\sum_{j=1}^{\infty}\psi_j(r_0)\,\psi_j(r),$$

which holds if

$$\frac{1}{\sin\theta}\frac{\mathrm{d}}{\mathrm{d}\theta}\left(\sin\theta\frac{\mathrm{d}\chi_j}{\mathrm{d}\theta}\right)-(\mu_j^2+\tfrac{1}{4})\,\chi_j=-\frac{\delta(\theta-\theta_0)}{2\pi\sin\theta},$$

that is to say,

$$\frac{1}{\sin\theta}\frac{\mathrm{d}}{\mathrm{d}\theta}\left(\sin\theta\frac{\mathrm{d}\chi_j}{\mathrm{d}\theta}\right)-(\mu_j^2+\tfrac{1}{4})\,\gamma_j=0\quad(\theta\neq\theta_0), \tag{6.6.15}$$

$$[\chi_j]_{\theta_0-0}^{\theta_0+0}=0,\quad\left[\sin\theta\frac{\mathrm{d}\chi_j}{\mathrm{d}\theta}\right]_{\theta_0-0}^{\theta_0+0}=-\frac{1}{2\pi}. \tag{6.6.16}$$

The solutions of (6.6.15) are Legendre functions of order $-\frac{1}{2}+\mathrm{i}\mu_j$. A solution which is finite at $\theta=0$ is

$$P_{-\frac{1}{2}+\mathrm{i}\mu_j}(\cos\theta)=F\left(+\tfrac{1}{2}+\mathrm{i}\mu_j,\tfrac{1}{2}-\mathrm{i}\mu_j;\ 1;\frac{1-\cos\theta}{2}\right). \tag{6.6.17}$$

It is real when μ_j is real. As a second independent solution we can take $P_{-\frac{1}{2}+\mathrm{i}\mu_j}(-\cos\theta)$; this is finite at $\theta=\pi$. It is a consequence of the results obtained in the appendix to chapter 3 that the Laplace transform of a pulse remains finite at a caustic. Since we are still dealing with a ring source ($\theta_0 \neq 0$) $\bar{G}$ must therefore be finite for $\theta = 0$, $\theta = \pi$. Hence

$$\chi_j = \begin{cases} KP_{-\frac{1}{2}+i\mu_j}(\cos\theta)\,P_{-\frac{1}{2}+i\mu_j}(-\cos\theta_0) & (0 \leqslant \theta \leqslant \theta_0), \\ KP_{-\frac{1}{2}+i\mu_j}(\cos\theta_0)\,P_{-\frac{1}{2}+i\mu_j}(-\cos\theta) & (\theta_0 \leqslant \theta \leqslant \pi), \end{cases} \quad (6.6.18)$$

where K is a constant. In order to determine it we must calculate the Wronskian of the two solutions of (6.6.15). Since†

$$P_\nu(-\cos\theta) = P_\nu(\cos\theta)\cos\pi\nu - \frac{\pi}{2}\sin\pi\nu Q_\nu(\cos\theta),$$

$$P'_\nu(\cos\theta)\,Q_\nu(\cos\theta) - P_\nu(\cos\theta)\,Q'_\nu(\cos\theta) = -\frac{1}{\sin^2\theta},$$

we have

$$P_\nu(\cos\theta)\frac{d}{d\theta}P_\nu(-\cos\theta) - P_\nu(-\cos\theta)\frac{d}{d\theta}P_\nu(\cos\theta) = \frac{2\sin\pi\nu}{\pi\sin\theta}.$$

Hence (6.6.18) will satisfy (6.6.16) if

$$K = -\frac{1}{4\sin\pi(-\frac{1}{2}+i\mu_j)} = \frac{1}{4\cosh\pi\mu_j}. \quad (6.6.19)$$

Substituting in (6.6.14) we obtain

$$\bar{G} = \frac{1}{4}\sum_{j=1}^{\infty}\psi_j(r_0)\,\psi_j(r)P_{-\frac{1}{2}+i\mu_j}(\cos\theta_0)\,P_{-\frac{1}{2}+i\mu_j}(-\cos\theta)\operatorname{sech}\pi\mu_j$$
$$(\theta_0 \leqslant \theta \leqslant \pi).$$

We can now deduce the Green's function due to a point source at $r = r_0$, $\theta = 0$ by making $\theta_0 \to 0$. If we also substitute (6.6.12) for the ψ_j we find that

$$\bar{G}(r,\theta,s;\,r_0) = \frac{1}{4(rr_0)^{\frac{1}{2}}}\sum_{j=1}^{\infty}\frac{L_j^2}{\cosh\pi\mu_j}K_{i\mu_j}\left(\frac{sr_0}{c}\right) \times K_{i\mu_j}\left(\frac{sr}{c}\right)P_{-\frac{1}{2}+i\mu_j}(-\cos\theta), \quad (6.6.20)$$

where the constants L_j are given by (6.6.13).

7. Approximate evaluation of the diffracted field

Following the same procedure as in the cylinder problem we now replace the series (6.6.20) by the asymptotic form of its leading

† See Hobson (1931), p. 231, eqn. (62); p. 233, eqn. (66).

term when s is large. For $\theta=\pi$ the Legendre functions reduce to unity and the formulae of §4 of the appendix are sufficient for this purpose. When $\theta \neq \pi$ the Legendre functions must also be approximated. The requisite formulae can be derived from the differential equation (6.6.15). For large μ_j and $\epsilon \leqslant \theta \leqslant \pi-\epsilon$, $\epsilon>0$, the solutions of this equation are asymptotically linear combinations of $(\sin\theta)^{-\frac{1}{2}}\,\mathrm{e}^{\pm\mu_j\theta}$. For small θ we can replace $\sin\theta$ by θ; (6.6.15) then reduces to the modified Bessel's equation of order zero, and since $P_\nu(0)=1$ we have

$$P_{-\frac{1}{2}+\mathrm{i}\mu_j}(\cos\theta) \sim I_0\{(\mu_j^2+1)^{\frac{1}{2}}\theta\} \sim (2\pi\mu_j\theta)^{-\frac{1}{2}}\,\mathrm{e}^{\mu_j\theta}.$$

Hence† $$P_{-\frac{1}{2}+\mathrm{i}\mu_j}(\cos\theta) \sim (2\pi\mu_j\sin\theta)^{-\frac{1}{2}}\,\mathrm{e}^{\mu_j\theta}. \tag{6.7.1}$$

Replacing θ by $\pi-\theta$ and multiplying by $\frac{1}{2}\operatorname{sech}\mu_j\pi \sim \mathrm{e}^{-\pi\mu_j}$ we obtain

$$\frac{P_{-\frac{1}{2}+\mathrm{i}\mu_j}(-\cos\theta)}{2\cosh\pi\mu_j} \sim \frac{\mathrm{e}^{-\mu_j\theta}}{(2\pi\mu_j\sin\theta)^{\frac{1}{2}}} = \frac{\mathrm{e}^{-\mu_j\theta}}{2\mu_j}\left(\frac{2\mu_j}{\pi\sin\theta}\right)^{\frac{1}{2}}. \tag{6.7.2}$$

For large s, the boundary condition (6.6.7) reduces to the equation (6.3.6) for the cylinder problem, and the eigenfunctions ψ_j given by (6.6.12) and (6.6.13) reduce to $r^{-\frac{1}{2}}\phi_j$, where ϕ_j is the function defined by (6.3.9). Hence the approximations to $\bar{G}$ are obtained by multiplying (6.4.1)–(6.4.3) (with $j=1$) by the factor

$$\frac{1}{2(rr_0)^{\frac{1}{2}}}\left(\frac{2\mu_1}{\pi\sin\theta}\right)^{\frac{1}{2}} \sim \left(\frac{sa}{2\pi crr_0\sin\theta}\right)^{\frac{1}{2}}.$$

By (6.4.7) this implies that the inverse Laplace transform must be multiplied by

$$\left(\frac{1}{2\pi rr_0\sin\theta}\right)^{\frac{1}{2}}\left(\frac{a}{c}\right)^{\frac{1}{2}}\left\{\left(\frac{a}{2c}\right)^{\frac{1}{3}}\frac{\alpha_1\delta}{3T}\right\}^{\frac{3}{4}} = \frac{\alpha_1^{\frac{3}{4}}}{6^{\frac{3}{4}}\pi^{\frac{1}{2}}}\left(\frac{a}{rr_0\sin\theta}\right)^{\frac{1}{2}}\delta^{\frac{3}{4}}\left(\frac{a}{cT}\right)^{\frac{3}{4}}.$$

† The formula (6.7.1) gives the leading term of the asymptotic expansion of $P_{-\frac{1}{2}+\mathrm{i}\mu_j}(\cos\theta)$ when $\mu_j\to\infty$. Since $P_{-\frac{1}{2}+\mathrm{i}\mu_j}(\cos\theta)=P_{-\frac{1}{2}-\mathrm{i}\mu_j}(\cos\theta)$, there must be a second term $\{2\pi|\mu_j|\sin\theta\}^{-\frac{1}{2}}\,\mathrm{e}^{-\mu_j\theta}$. The complete expansion of (6.7.2) therefore contains terms in $\exp\{-\mu_j(\theta+2m\pi)\}$ $(m=0, 1, 2, \ldots)$ and in $\exp\{-\mu_j(2m\pi-\theta)\}$ $(m=1, 2, \ldots)$, each multiplied by an asymptotic series in descending powers of $\mu_j^{\frac{1}{3}}$. Each of these terms when multiplied by the appropriate r-eigenfunction represents a pulse-mode that has encircled the sphere a number of times, either in the sense of increasing θ or in the sense of decreasing θ. Just as in the problem of the circular cylinder, our approximations permit us only to take the first pulse which arrives into account, and for this purpose only the approximations (6.7.1) and (6.7.2) are required.

Hence we can deduce from (6.4.8)–(6.4.10) that

$$G(r,\theta,t;\,r_0) \sim \frac{\alpha_1^{\frac{3}{4}} A}{6^{\frac{3}{4}}\pi^{\frac{1}{2}}}\frac{c}{a^2}\frac{a}{(rr_0\sin\theta)^{\frac{1}{2}}} \times \left[\frac{a^4}{(r^2-a^2)(r_0^2-a^2)}\right]^{\frac{1}{4}} \delta^{\frac{1}{2}}\frac{a}{cT}H(T)\,\mathrm{e}^{-\xi}, \quad (6.7.3)$$

$$G(a,\theta,t;\,r_0) \sim \frac{\alpha_1^{\frac{3}{4}} B}{6^{\frac{3}{4}}\pi^{\frac{1}{2}}}\frac{c}{a^2}\left(\frac{a}{r_0\sin\theta}\right)^{\frac{1}{2}} \times \left[\frac{a^2}{r_0^2-a^2}\right]^{\frac{1}{4}} \delta^{\frac{3}{4}}\left(\frac{a}{cT}\right)^{\frac{5}{4}} H(T)\,\mathrm{e}^{-\xi}, \quad (6.7.4)$$

$$G(a,\theta,t;\,a) \sim \frac{\alpha_1^{\frac{3}{4}} C}{6^{\frac{3}{4}}\pi^{\frac{1}{2}}}\frac{c}{a^2}\frac{1}{(\sin\theta)^{\frac{1}{2}}}\,\delta\left(\frac{a}{cT}\right)^{\frac{3}{2}} H(T)\,\mathrm{e}^{-\xi}, \quad (6.7.5)$$

where ξ is again defined by (6.4.11) and the constants A, B and C are given by (6.4.12). These diffraction formulae hold for $\theta \neq \pi$.

When $\theta = \pi$ the series (6.6.20) becomes

$$\bar{G}(r,\pi,s;\,r_0) = \frac{1}{4(rr_0)^{\frac{1}{2}}}\sum_{j=1}^{\infty}\frac{L_j K_{\mathrm{i}\mu_j}(sr_0/c)\,K_{\mathrm{i}\mu_j}(sr/c)}{\cosh\pi\mu_j}. \quad (6.7.6)$$

If we retain only the first mode and replace $\operatorname{sech}\pi\mu_1$ by $2\mathrm{e}^{-\pi\mu_1}$ and the numerator by its asymptotic approximation we obtain (6.4.1)–(6.4.3) with $j = 1$, multiplied by

$$\mu_1(rr_0)^{-\frac{1}{2}} \sim \frac{sa}{c(rr_0)^{\frac{1}{2}}}.$$

By (6.4.7) this implies that the inverse Laplace transform must be multiplied by

$$\frac{a}{c(rr_0)^{\frac{1}{2}}}\left\{\left(\frac{a}{2c}\right)^{\frac{1}{3}}\frac{\alpha_1\delta}{3T}\right\}^{\frac{3}{2}} = \frac{1}{a}\frac{2\alpha_1^{\frac{3}{2}}}{6^{\frac{3}{2}}}\left(\frac{a}{rr_0}\right)^{\frac{1}{2}}\delta^{\frac{3}{2}}\left(\frac{a}{cT}\right)^{\frac{3}{2}}.$$

Hence

$$G(r,\pi,t;\,r_0) \sim \frac{2\alpha_1^{\frac{3}{2}} A}{6^{\frac{3}{2}}}\frac{c}{a^2}\left(\frac{a}{rr_0}\right)^{\frac{1}{2}}\left[\frac{a^4}{(r^2-a^2)(r_0^2-a^2)}\right]^{\frac{1}{4}} \times \delta^{\frac{5}{4}}\left(\frac{a}{cT}\right)^{\frac{7}{4}} H(T)\mathrm{e}^{-\xi}, \quad (6.7.7)$$

$$G(a,\pi,t;\,r_0) \sim \frac{2\alpha_1^{\frac{3}{2}} B}{6^{\frac{3}{2}}}\frac{c}{a^2}\left(\frac{a}{r_0}\right)^{\frac{1}{2}}\left(\frac{a^2}{r_0^2-a^2}\right)^{\frac{1}{4}}\delta^{\frac{3}{2}}\left(\frac{a}{cT}\right)^{2} H(T)\,\mathrm{e}^{-\xi}, \quad (6.7.8)$$

$$G(a,\pi,t;\,a) \sim \frac{2\alpha_1^{\frac{3}{2}} C}{6^{\frac{3}{2}}}\frac{c}{a^2}\delta^{\frac{7}{4}}\left(\frac{a}{cT}\right)^{\frac{9}{4}} H(T)\,\mathrm{e}^{-\xi}. \quad (6.7.9)$$

These diffraction formulae give the Green's function which is the diffracted field of the pulse

$$E=\frac{\delta(t-R/c)}{4\pi R}, \quad R=(r^2+r_0^2-2rr_0\cos\theta)^{\frac{1}{2}}. \qquad (6.7.10)$$

A case of more immediate physical interest is the diffraction of an acoustic spherical shock wave, say

$$p_1=\frac{H(ct-R)}{R}. \qquad (6.7.11)$$

Since this is the integral of $4\pi E$ with respect to t, the diffraction formulae can be derived by multiplying the approximate expressions for G by $4\pi/s$ and taking the inverse Laplace transform. To our order of approximation it follows from (6.4.7) that this is equivalent to multiplying the diffraction formulae for G by

$$4\pi\left[\left(\frac{a}{2c}\right)^{\frac{1}{3}}\frac{\alpha_1\delta}{3T}\right]^{-\frac{3}{2}}=2\pi\frac{6^{\frac{1}{2}}}{\alpha_1^{\frac{3}{2}}}\delta^{-\frac{3}{2}}\left(\frac{cT}{a}\right)^{\frac{3}{2}}\frac{a}{c}, \qquad (6.7.12)$$

and there is no need to write the resulting approximations down in detail.

Finally, we can deduce the diffracted field due to an incident plane unit pressure pulse

$$p_1=H(ct+r\cos\theta). \qquad (6.7.13)$$

All we need to do is to multiply the diffraction formulae by the factor (6.7.12) and by r_0, to replace T by $T+r_0/c$ and to make $r_0\to\infty$. The following approximations are then obtained:

$$P(r,\theta,t)\sim A'\frac{a}{(r\sin\theta)^{\frac{1}{2}}(r^2-a^2)^{\frac{1}{4}}}\frac{1}{\gamma}\left(\frac{cT}{a}\right)^{\frac{1}{2}}H(T)\,\mathrm{e}^{-\xi}, \qquad (6.7.14)$$

$$P(a,\theta,t)\sim B'\frac{1}{(\sin\theta)^{\frac{1}{2}}}\gamma^{-\frac{3}{4}}\left(\frac{cT}{a}\right)^{\frac{1}{4}}H(T)\,\mathrm{e}^{-\xi}, \qquad (6.7.15)$$

$$P(r,\pi,t)\sim 4\pi A\frac{a}{r^{\frac{1}{2}}(r^2-a^2)^{\frac{1}{4}}}\gamma^{-\frac{1}{4}}\left(\frac{a}{cT}\right)^{\frac{1}{4}}H(T)\,\mathrm{e}^{-\xi}, \qquad (6.7.16)$$

$$P(a,\pi,t)\sim 4\pi B\left(\frac{a}{cT}\right)^{\frac{1}{2}}H(T)\,\mathrm{e}^{-\xi}. \qquad (6.7.17)$$

Here A' and B' are the same constants as those which figure in the two-dimensional case, (6.5.15). Also, $T=t-\gamma$, and γ, σ are defined by (6.5.8) and (6.5.9). In fact, the diffracted field in the shadow of the sphere differs from that in the shadow of the cylinder (6.5.13) and (6.5.14) only by the additional divergence factor $(a/r\sin\theta)^{\frac{1}{2}}$. The pressure therefore rises more quickly, as one would expect. But on the axis in the shadow ($\theta=\pi$) it is given by (6.7.16) and (6.7.17) and is strikingly different from the two-dimensional case. This is illustrated by the curves in figs. 6.2 and 6.3 which refer to the pressure field on the sphere. In fig. 6.3 which shows P as a function of $\gamma=\theta-\frac{1}{2}\pi$ for constant cT/a each curve has a sharp peak at $\gamma=90°$ which becomes more pronounced as T increases. Fig. 6.2 shows that the pressure at $\gamma=90°$ is roughly in the same range as that at $\gamma=50°$ in the interval $0\leqslant cT/a\leqslant 0{\cdot}02$. This is of course a focusing effect, and in the harmonic case a well-known optical phenomenon.

8. Geometrical optics in a stratified medium

In a homogeneous medium shadows are cast by curved obstacles; in an inhomogeneous one they may arise because of the curvature of the rays. This may be illustrated by considering the geometrical optics of a stratified medium. Suppose that the velocity of sound depends on one space coordinate, say z, only. Then the eikonal equation is

$$\left(\frac{\partial\tau}{\partial x}\right)^2+\left(\frac{\partial\tau}{\partial y}\right)^2+\left(\frac{\partial\tau}{\partial z}\right)^2=g(z), \quad g=\frac{1}{c^2}. \tag{6.8.1}$$

The differential equations (2.3.7) of the rays are

$$\left.\begin{aligned} &\frac{dx}{d\nu}=\frac{\partial\tau}{\partial x}, \quad \frac{dy}{d\nu}=\frac{\partial\tau}{\partial y}, \quad \frac{dz}{d\nu}=\frac{\partial\tau}{\partial z}, \\ &\frac{d}{d\nu}\left(\frac{\partial\tau}{\partial x}\right)=0, \quad \frac{d}{d\nu}\left(\frac{\partial\tau}{\partial y}\right)=0, \quad \frac{d}{d\nu}\left(\frac{\partial\tau}{\partial z}\right)=\tfrac{1}{2}g'(z). \end{aligned}\right\} \tag{6.8.2}$$

Hence
$$\frac{dx}{d\nu}=\frac{\partial\tau}{\partial x}=\lambda\cos\alpha, \quad \frac{dy}{d\nu}=\frac{\partial\tau}{\partial y}=\lambda\sin\alpha, \tag{6.8.3}$$

where λ and α are constants, and so by (6.8.1) and (6.8.2)

$$\left(\frac{dz}{d\nu}\right)^2=g(z)-\lambda^2. \tag{6.8.4}$$

This gives two values for $dz/d\nu$ of opposite sign. Consider an arc of a ray on which $g(z)-\lambda^2$ does not change sign; let (x_1, y_1, z_1) and (x, y, z) be its end-points and suppose that $z > z_1$. Then

$$x - x_1 = \lambda\nu\cos\alpha, \quad y - y_1 = \lambda\nu\sin\alpha, \quad \nu = \int_{z_1}^{z} \frac{dz'}{[g(z')-\lambda^2]^{\frac{1}{2}}}. \tag{6.8.5}$$

The rays are therefore in planes parallel to the z-axis. The geodesic length τ of this arc can be calculated as follows. By (6.8.1) and (6.8.2),

$$\frac{d\tau}{d\nu} = g(z),$$

whence

$$\tau(x, y, z) - \tau(x_1, y_1, z_1) = \int_0^{\nu} g(z')\, d\nu', \quad \nu' = \int_{z_1}^{z'} \frac{dz''}{[g(z'')-\lambda^2]^{\frac{1}{2}}}.$$

Hence

$$\begin{aligned}\tau(x, y, z) - \tau(x_1, y_1, z_1) &= \int_0^{\nu} [g(\tau')-\lambda^2]\, d\nu' + \lambda^2\nu \\ &= \int_{z_1}^{z} [g(\tau')-\lambda^2]^{\frac{1}{2}}\, dz' + \lambda^2\nu,\end{aligned}$$

or, by (6.8.5),

$$\begin{aligned}\tau(x, y, z) - \tau(x_1, y_1, z_1) = \lambda[(x-x_1)^2 + (y-y_1)^2]^{\frac{1}{2}} \\ + \int_{z_1}^{z} [g(z')-\lambda^2]^{\frac{1}{2}}\, dz'.\end{aligned} \tag{6.8.6}$$

If (x_1, y_1, z_1) and (x, y, z) are given then λ is determined by (6.8.5). But these equations imply that the derivative of the right-hand side of (6.8.6) with respect to λ is zero, so that λ can also be determined by the condition that it makes this right-hand side stationary.

Let us now suppose that the medium fills the half-space $z \geqslant 0$ and that the rays are reflected at the boundary $z = 0$, and consider the rays that issue from the point $S(0, 0, z_0)$, where $z_0 > 0$. Since there is axial symmetry, we need only discuss the rays in a fixed plane that contains the z-axis, say the plane $y = 0$. We will assume that $g(z)$ is an increasing function of z that tends to infinity with z. By (6.8.4), only rays with $\lambda^2 < g(z_0)$ are admissible since $dz/d\nu$ must be real at S. To every admissible λ there correspond two rays, one with $(dz/dx)_s > 0$ and the other with $(dz/dx)_s < 0$. On the first ray z increases steadily, and its equation is

$$x = \int_{z_0}^{z} \frac{\lambda\, dz'}{[g(z')-\lambda^2]^{\frac{1}{2}}}. \tag{6.8.7}$$

by (6.8.5). On the second ray, z decreases to a minimum z^* such that

$$g(z^*) = \lambda^2, \tag{6.8.8}$$

and then increases again. There are now two possibilities. If $g(0) < \lambda^2$ so that $z^* > 0$ then the ray does not meet the boundary but turns away from it and continues towards infinite z. But if $\lambda^2 < g(0)$ then the ray meets the boundary and is reflected there. The transitional case is the 'glancing ray' for which $\lambda^2 = g(0)$.

In the first case, the ray has a 'descending' branch

$$x = \int_z^{z_0} \frac{\lambda\,dz'}{[g(z') - \lambda^2]^{\frac{1}{2}}} \quad (z^* \leqslant z \leqslant z_0). \tag{6.8.9}$$

The 'rising' branch is

$$x = x^* + \lambda'\nu, \quad \nu = \int_{z^*}^{z} \frac{dz'}{[g(z') - \lambda'^2]^{\frac{1}{2}}},$$

where x^* is the abscissa of the lowest point, where $z = z^*$, and

$$\lambda' = \left\{\left[g(z) - \left(\frac{dz}{d\nu}\right)^2\right]_{z=z^*}\right\}^{\frac{1}{2}} = [g(z^*)]^{\frac{1}{2}} = \lambda.$$

Hence the rising branch has the equation

$$x = \int_{z^*}^{z_0} \frac{\lambda\,dz'}{[g(z') - \lambda^2]^{\frac{1}{2}}} + \int_{z^*}^{z} \frac{\lambda\,dz'}{[g(z') - \lambda^2]^{\frac{1}{2}}}. \tag{6.8.10}$$

For the glancing ray this equation becomes

$$x = x_0 + \int_0^z \left\{\frac{g(0)}{g(z') - g(0)}\right\}^{\frac{1}{2}} dz', \quad x_0 = \int_0^{z_0} \left\{\frac{g(0)}{g(z') - g(0)}\right\}^{\frac{1}{2}} dz'. \tag{6.8.11}$$

An incident ray for which $\lambda^2 < g(0)$ has the equation (6.8.9) and meets the boundary when $x = x'$, where

$$x' = \int_0^{z_0} \frac{\lambda\,dz'}{[g(z') - \lambda^2]^{\frac{1}{2}}}. \tag{6.8.12}$$

The reflected ray issues from $(x', 0, 0)$ with a slope equal in magnitude but opposite in sign to that of the incident ray. Hence λ is unchanged and the equation of the reflected ray is

$$x = \int_0^{z_0} \frac{\lambda\,dz'}{[g(z') - \lambda^2]^{\frac{1}{2}}} + \int_0^{z} \frac{\lambda\,dz'}{[g(z') - \lambda^2]^{\frac{1}{2}}}. \tag{6.8.13}$$

We can now show that the reflected ray is in the domain bounded by the z-axis, the segment $(0, x_0)$ of the boundary, and the rising branch of the glancing ray. For if $\bar{x}$ refers to the rising branch of the glancing ray and x to the reflected ray then by (6.8.11) and (6.8.13)

$$\frac{d}{dz}(\bar{x}-x)=\left\{\frac{g(0)}{g(z)-g(0)}\right\}^{\frac{1}{2}}-\frac{\lambda}{[g(z)-\lambda^2]^{\frac{1}{2}}}>0,$$

since
$$\frac{g(0)}{g(z)-g(0)}-\frac{\lambda^2}{g(z)-\lambda^2}=\frac{g(z)[g(0)-\lambda^2]}{[g(z)-g(0)][g(z)-\lambda^2]},$$

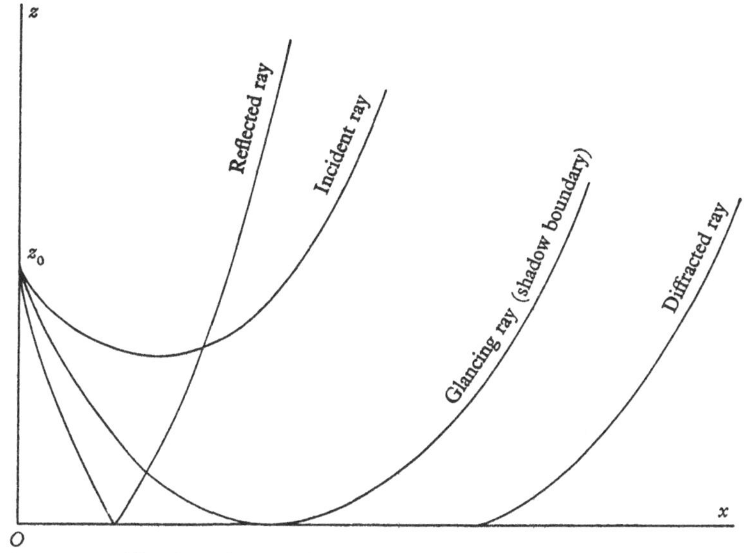

Fig. 6.4. Geometrical optics in a stratified medium.

and $\lambda^2 < g(0)$ by hypothesis. Since also $[\bar{x}-x]_{z=0}=x_0-x'>0$, it follows that $\bar{x}-x>0$ for all $z \geqslant 0$, which proves the assertion. Hence no reflected ray penetrates into the domain bounded by $0 \leqslant x \leqslant x_0$, $z=0$ and the rising branch of the glancing ray. This domain is therefore a shadow (fig. 6.4). There are cases where incident rays enter the shadow and are turned back at a caustic; this is a complication which we shall ignore.

Through every point of the shadow there is a diffracted ray which is tangential to the boundary. If $(x', 0, 0)$ is the foot of a diffracted ray then the equation of the ray is, by (6.8.5),

$$x=x'+\int_0^z\left\{\frac{g(0)}{g(0)-g(z')}\right\}^{\frac{1}{2}}dz', \qquad (6.8.14)$$

for as the ray touches $z=0$ we must have $\lambda=[g(0)]^{\frac{1}{2}}$. When x and z are given this equation determines x'. The diffracted front arrives at (x, z) when $t=\tau$, where τ is the geodesic distance of (x, z) from $(0, z_0)$ measured along the descending branch of the glancing ray, the segment (x_0, x') of the boundary, and the diffracted ray. By (6.8.6)

$$\tau=\left\{\int_0^{z_0}[g(z')-g(0)]^{\frac{1}{2}}\,dz'+[g(0)]^{\frac{1}{2}}x_0\right\}+[g(0)]^{\frac{1}{2}}(x'-x_0)$$
$$+\left\{\int_0^{z}[g(z')-g(0)]^{\frac{1}{2}}\,dz'+[g(0)]^{\frac{1}{2}}(x-x')\right\},$$

whence

$$\tau=[g(0)]^{\frac{1}{2}}x+\int_0^{z_0}[g(z')-g(0)]^{\frac{1}{2}}\,dz'+\int_0^{z}[g(z')-g(0)]^{\frac{1}{2}}\,dz'.$$

At a point not in the xz-plane we must replace x by $\varpi+(x^2+y^2)^{\frac{1}{2}}$, so that

$$\tau=[g(0)]^{\frac{1}{2}}\varpi+\int_0^{z_0}[g(z')-g(0)]^{\frac{1}{2}}\,dz'+\int_0^{z}[g(z')-g(0)]^{\frac{1}{2}}\,dz'. \quad (6.8.15)$$

The 'depth' of a point in the shadow can be specified by giving the distance between the foot of the diffracted ray through the point from the shadow boundary, $x'-x_0$. By (6.8.11) and (6.8.14)

$$\delta=x'-x_0=x-\int_0^{z}\left\{\frac{g(0)}{g(z')-g(0)}\right\}^{\frac{1}{2}}dz'-\int_0^{z_0}\left\{\frac{g(0)}{g(z')-g(0)}\right\}^{\frac{1}{2}}dz',$$

and so in the general case

$$\delta=\varpi-\int_0^{z}\left\{\frac{g(0)}{g(z')-g(0)}\right\}^{\frac{1}{2}}dz'-\int_0^{z_0}\left\{\frac{g(0)}{g(z')-g(0)}\right\}^{\frac{1}{2}}dz'. \quad (6.8.16)$$

In the corresponding two-dimensional case the source is a line source parallel to the x-axis and at distance z_0 from it. The rays are then parallel to the xy-plane. The formulae (6.8.15) and (6.8.16) can be adapted to this case by writing $|x|$ instead of ϖ.

9. Pulse diffraction in a stratified medium

The results of the preceding section show that the pulse due to a point or line source, that is to say, the Green's function, in a stratified medium with a plane boundary, is subject to diffraction. The diffracted field can again be approximated by diffraction formulae of the type (6.1.1). To simplify the exposition we shall confine ourselves here to the equations

$$\frac{1}{c^2}\frac{\partial^2 G}{\partial t^2} - \nabla^2 G = \delta(x)\,\delta(y)\,\delta(z-z_0)\,\delta(t) \quad (z>0,\ z_0>0), \qquad (6.9.1)$$

and

$$\frac{1}{c^2}\frac{\partial^2 G}{\partial t^2} - \nabla^2 G = \delta(x)\,\delta(z-z_0)\,\delta(t) \qquad (6.9.2)$$

for the three- and two-dimensional cases respectively, with the boundary condition

$$\left[\frac{\partial G}{\partial z}\right]_{z=0} = 0, \qquad (6.9.3)$$

appropriate to a rigid fixed boundary. Also, G is compact towards the past. According to the theory of § 1.4 this equation is satisfied by short pressure pulses in a medium in convective equilibrium subject to a field of force in the z-direction that depends on z only, for instance, a uniform gravitational field.† It is to be noted that other problems can be brought into this form. For instance, the substitution $r = a\,e^z$ transforms the two-dimensional wave equation in polar coordinates into

$$\frac{\partial^2 p}{\partial \theta^2} + \frac{\partial^2 p}{\partial z^2} = \frac{a^2}{c^2}\,e^{2z}\,\frac{\partial^2 p}{\partial t^2},$$

and the boundary condition $[\partial p/\partial r]_{r=a} = 0$ into $[\partial p/\partial z]_{z=0} = 0$; hence the cylinder problem which we have already discussed is equivalent to a stratified medium problem.‡

We begin again by introducing the 'eigenfunctions in z' which satisfy the differential equation

$$\phi''(z) + [\mu^2 - s^2 g(z)]\,\phi(z) = 0, \quad g = \frac{1}{c^2}, \qquad (6.9.4)$$

† In an ideal gas one has then $c^2 = c_0^2 - (\gamma - 1)\,gz$, where c_0 is the velocity of sound at ground level, g the gravitational acceleration, and γ the ratio of the specific heats at constant pressure and volume respectively. Taking $g = 981$ cm./sec.^{2} and $c_0 = 330$ m./sec., $\gamma = 1{\cdot}4$ for air one finds that $c_0^2/(\gamma-1)\,g = 26{\cdot}7$ km. The actual problem of pulse propagation in the atmosphere is complicated by humidity which affects the temperature gradient, by the effect of wind, and by the scattering of sound due to turbulence.

‡ Diffraction by elliptic and parabolic cylinders can be considered similarly and leads to the wave equation (6.9.2) with $c^{-2} = f(x) + g(z)$. This can be treated in the same way as the problem considered here, and again leads to diffraction formulae of the type (6.1.1) for the field near the diffracted front (Friedlander, 1955).

and the boundary condition

$$\phi'(0) = 0. \tag{6.9.5}$$

They must also tend to zero as $z \to \infty$. Since by hypothesis $g(z) \to \infty$ as $z \to \infty$ there is (for real s) a discrete spectrum of real positive eigenvalues μ_j ($j = 1, 2, \ldots$). The corresponding eigenfunctions ϕ_j can be normalized so that

$$\int_0^\infty \phi_j \phi_k \, dz = \delta_{jk}, \tag{6.9.6}$$

where the δ_{jk} are Kronecker deltas. The formal completeness relation is then

$$\delta(z - z_0) = \sum_{j=1}^{\infty} \phi_j(z_0)\, \phi_j(z). \tag{6.9.7}$$

It will be valid in the sense of the theory of distributions if a suitable condition on the growth of g as $z \to \infty$ is imposed. Hence the Laplace transforms of the equations (6.9.1) and (6.9.2) will be satisfied by

$$\bar{G} - \sum_{j=1}^{\infty} \phi_j(z_0)\, \phi_j(z)\, \chi_j,$$

provided that $$\frac{\partial^2 \chi_j}{\partial x^2} + \frac{\partial^2 \chi_j}{\partial y^2} - \mu_j^2 \chi_j = -\delta(x)\,\delta(y)$$

in the three-dimensional case and

$$\frac{d^2 \chi_j}{dx^2} - \mu_j^2 \chi_j = -\delta(x)$$

in the two-dimensional case. The χ_j must remain finite as $\varpi \to \infty$ or $|x| \to \infty$ respectively. Hence by (5.A.21)

$$\bar{G} = \frac{1}{2\pi} \sum_{j=1}^{\infty} \phi_j(z_0)\, \phi_j(z)\, K_0(\mu_j \varpi) \tag{6.9.8}$$

in the three-dimensional case, and (just as in the cylinder problem)

$$\bar{G} = \sum_{j=1}^{\infty} \frac{\phi_j(z_0)\, \phi_j(z)}{2\mu_j} e^{-\mu_j |x|} \tag{6.9.9}$$

in the two-dimensional case. We assume that these expansions are also valid for complex s, at least in $\mathscr{R}s > 0$. Examples show that this is not necessarily true in the whole of the shadow; a case in point is the simple linear law $g = 1 + z$, where it can be shown that the series converges in $\mathscr{R}s > 0$ only if $\varpi > (z + z_0)\sqrt{3}$ (Friedlander, 1955, appendix II). It is possible that the diffraction formulae still give the field near the diffracted front but must be established by a different method; one could probably start with the contour integral representation of $\bar{G}$ which is equivalent to the series expansion when this converges but is valid everywhere except on $\varpi = 0$.

It remains to determine the behaviour of the μ_j as $s \to \infty$ and to derive asymptotic formulae for the ϕ_j. This can be done by Langer's method as in the problems which we have already considered. Alternatively, one can use the following formal method which is simpler. Let

$$z = g_1^{-\frac{1}{3}} s^{-\frac{2}{3}} \zeta, \quad \mu^2 = s^2 g_0 + \alpha g_1^{-\frac{2}{3}} s^{-\frac{4}{3}}, \quad g_0 = g(0), \quad g_1 = g'(0). \quad (6.9.10)$$

It is assumed that $g'(0) > 0$, which is the general case. Then the differential equation (6.9.4) becomes

$$\frac{d^2\phi}{d\zeta^2} + [\alpha - \zeta - \epsilon q(\zeta, \epsilon)]\, \phi = 0, \quad (6.9.11)$$

where $$\epsilon = g_1^{-\frac{1}{3}} s^{-\frac{2}{3}}$$

and $$q = \frac{g(\epsilon\zeta) - g_0 - \epsilon g_1 \zeta}{\epsilon^2 g_1}.$$

Since q remains bounded as $\epsilon \to 0$ the term ϵq in (6.9.11) is small when s is large. We may therefore expect that an approximation to ϕ valid for large s is obtained when this term is ignored. Thus

$$\phi \sim L\,\mathrm{Ai}\,(\zeta - \alpha), \quad (6.9.12)$$

where L is a constant. Hence the boundary condition (6.9.5) implies that $\alpha \sim \alpha_j$, where $-\alpha_1, -\alpha_2, \ldots$ are again the zeros of $\mathrm{Ai}'(\alpha)$, and so

$$\mu_j^2 \sim s^2 g_0 + \alpha_j g_1^{\frac{2}{3}} s^{\frac{4}{3}}. \quad (6.9.13)$$

The argument can be extended formally to yield an expansion of μ_j in descending powers of $s^{\frac{2}{3}}$.† Hence

$$\phi_j \sim \psi_j = L_j\,\mathrm{Ai}(g_1^{\frac{1}{3}} s^{\frac{2}{3}} z - \alpha_j) \quad (j = 1, 2, \ldots).$$

Now $$\psi_j'' + (\alpha_j g_1^{\frac{2}{3}} s^{\frac{4}{3}} - g_1 s^2 z)\, \psi_j = 0,$$

whence $$\frac{d}{dz}\{(g_1 s^2 z - \alpha_j g_1^{\frac{2}{3}} s^{\frac{4}{3}})\, \psi_j^2 - \psi_j'^2\} = g_1 s^2 \psi_j^2.$$

† Friedlander (1955), appendix 1.

As the eigenfunctions are normalized, we have

$$1 \sim \int_0^\infty \psi_j^2 \, \mathrm{d}z = \alpha_j g_1^{-\frac{1}{3}} s^{-\frac{2}{3}} [\psi_j(0)]^2.$$

Hence when $s^{\frac{2}{3}}z$ is small

$$\phi_j \sim \frac{g_1^{\frac{1}{6}} s^{\frac{1}{3}}}{\alpha_j^{\frac{1}{2}} \operatorname{Ai}(-\alpha_j)} \operatorname{Ai}(g_1^{\frac{1}{3}} s^{\frac{2}{3}} z - \alpha_j). \tag{6.9.14}$$

We also require an asymptotic formula for ϕ_j when $z > 0$ and $s^{\frac{2}{3}}z$ is not small. In order to derive this we write (6.9.4) as

$$\frac{\mathrm{d}}{\mathrm{d}z}\left(\frac{\phi_j'}{\phi_j}\right) + \left(\frac{\phi_j'}{\phi_j}\right)^2 = s^2 g - \mu_j^2,$$

whence, by (6.9.13), supplemented by an error term,

$$\frac{\mathrm{d}}{\mathrm{d}z}\left(\frac{\phi_j'}{\phi_j}\right) + \left(\frac{\phi_j'}{\phi_j}\right)^2 = s^2(g - g_0) - \alpha_j g_1^{\frac{2}{3}} s^{\frac{4}{3}} + O(s^{\frac{2}{3}}).$$

Now let $$\frac{\phi_j'}{\phi_j} = -a(z)\, s + b(z)\, s^{\frac{1}{3}} + A(z) + O(s^{-\frac{1}{3}}).$$

Then

$$s^2 a^2 - 2ab s^{\frac{4}{3}} - (2aA + a')\, s + O(s^{\frac{2}{3}}) = s^2(g - g_0) - \alpha_j g_1^{\frac{2}{3}} s^{\frac{4}{3}} + O(s^{\frac{2}{3}}).$$

Hence $$a = (g - g_0)^{\frac{1}{2}}, \quad b = \tfrac{1}{2}\alpha_j g_1^{\frac{2}{3}} (g - g_0)^{-\frac{1}{2}}, \quad B = -\frac{a'}{2a},$$

and so

$$\phi_j \sim \frac{M}{(g - g_0)^{\frac{1}{4}}} \exp\left\{ -s \int_0^z [g(z') - g_0]^{\frac{1}{2}} + \tfrac{1}{2}\alpha_j\, g_1^{\frac{2}{3}} s^{\frac{1}{3}} \int_0^z \frac{\mathrm{d}z'}{[g(z') - g_0]^{\frac{1}{2}}} \right\}, \tag{6.9.15}$$

where M is independent of z. For small z this becomes

$$\phi_j \sim \frac{M}{g_1^{\frac{1}{4}} z^{\frac{1}{4}}} \exp\left(-\tfrac{2}{3} g_1^{\frac{1}{2}} s z^{\frac{3}{2}} + \alpha_j\, g_1^{\frac{1}{6}} s^{\frac{1}{3}} z^{\frac{1}{2}}\right).$$

But if the asymptotic formula (6.A.30) is substituted for the Airy function in (6.9.14) one finds that

$$\phi_j \sim \frac{g_1^{\frac{1}{6}} s^{\frac{1}{3}}}{\alpha_j^{\frac{1}{2}} \operatorname{Ai}(-\alpha_j)} \frac{1}{2\pi^{\frac{1}{2}} g_1^{\frac{1}{12}} s^{\frac{1}{6}} z^{\frac{1}{4}}} \exp\left(-\tfrac{2}{3} g_1^{\frac{1}{2}} s z^{\frac{3}{2}} + \alpha_j\, g_1^{\frac{1}{6}} s^{\frac{1}{3}} z^{\frac{1}{2}}\right).$$

Comparison of this with the preceding equation gives the value of M, and when this is substituted in (6.9.15) one obtains

$$\phi_j(z) \sim \frac{g_1^{\frac{1}{3}} s^{\frac{1}{6}}}{2(\pi\alpha_j)^{\frac{1}{2}} \operatorname{Ai}(-\alpha_j)} \frac{1}{[g(z) - g_0]^{\frac{1}{4}}}$$
$$\times \exp\left\{ -s \int_0^z [g(z') - g_0]^{\frac{1}{2}} \, \mathrm{d}z' + \tfrac{1}{2}\alpha_j\, g_1^{\frac{2}{3}} s^{\frac{1}{3}} \int_0^z \frac{\mathrm{d}z'}{[g(z') - g_0]^{\frac{1}{2}}} \right\}. \tag{6.9.16}$$

This is the approximation that must be substituted in (6.9.8) and (6.9.9) when $z>0$ or $z_0>0$; for $z=0$ or $z_0=0$ one must use the approximation derived from (6.9.14),

$$\phi_j(0) \sim \frac{g_1^{\frac{1}{6}}}{\alpha_j^{\frac{1}{2}}} s^{\frac{1}{2}}. \tag{6.9.17}$$

Finally,

$$K_0(\mu_j \varpi) \sim \left(\frac{\pi}{2\mu_j \varpi}\right)^{\frac{1}{2}} e^{-\mu_j \varpi} \sim \left(\frac{\pi}{2\varpi s}\right)^{\frac{1}{2}} \frac{1}{g_0^{\frac{1}{4}}} \exp\left(-\varpi g_0^{\frac{1}{2}} s - \tfrac{1}{2}\varpi g_1^{\frac{2}{3}} g_0^{-\frac{1}{2}} \alpha_j s^{\frac{1}{3}}\right). \tag{6.9.18}$$

When the approximations (6.9.16)–(6.9.18) are substituted in (6.9.8) it is found that each term of the eigenfunction expansion contains the factor

$$\exp\{-s\tau - \tfrac{1}{2}\alpha_j g_1^{\frac{2}{3}} g_0^{-\frac{1}{2}} \delta s^{\frac{1}{3}}\},$$

where τ is the arrival time of the diffracted pulse, (6.8.15), and δ is the quantity defined by (6.8.16) which is positive in the shadow. The situation is similar to that in the cases of the cylinder and the sphere which we have already considered. Following the procedure adopted in these cases we therefore reject all modes except the first one, $j=1$, and derive the diffracted fields by means of (6.4.7). This calculation can be arranged as follows. When $z=z_0=0$, the eigenfunctions must be approximated by (6.9.17) and we obtain

$$\bar{G} \sim \frac{1}{2\pi}\left(\frac{\pi}{2}\right)^{\frac{1}{2}} \frac{g_1^{\frac{1}{3}}}{\alpha_1 g_0^{\frac{1}{4}} \varpi^{\frac{1}{2}}} s^{\frac{1}{6}} \exp\{-s\tau - \tfrac{1}{2}\alpha_1 g_0^{-\frac{1}{2}} g_1^{\frac{2}{3}} \delta s^{\frac{1}{3}}\}, \tag{6.9.19}$$

whence

$$G \sim A\varpi^{-\frac{1}{2}} \delta T^{-\frac{3}{2}} H(T) e^{-\xi} \quad (z=z_0=0), \tag{6.9.20}$$

where

$$\xi = \frac{1}{3\sqrt{6}} \frac{g_1}{g_0^{\frac{3}{4}}} \alpha_1^{\frac{3}{2}} \frac{\delta^{\frac{3}{2}}}{T^{\frac{1}{2}}}, \quad A = \frac{1}{8\pi\sqrt{6}} \frac{g_1}{g_0^{\frac{3}{4}}}. \tag{6.9.21}$$

If next $z_0=0$, $z>0$, then (6.9.16) must be used for $\phi_1(z)$ instead of (6.9.17). Thus (6.9.19) must be multiplied by

$$\frac{g_1^{\frac{1}{6}} s^{-\frac{1}{6}}}{2\pi^{\frac{1}{2}} \operatorname{Ai}(-\alpha_1)} \frac{1}{[g(z)-g_0]^{\frac{1}{4}}}.$$

By (6.4.7) the effect of this on the inverse Laplace transform is, to our order of approximation, to multiply it by $D(z)\,\delta^{-\frac{1}{4}} T^{\frac{1}{4}}$, where

$$D(z) = \frac{6^{\frac{1}{4}} g_0^{\frac{1}{8}}}{2\pi^{\frac{1}{2}} \alpha_1^{\frac{1}{4}} \operatorname{Ai}(-\alpha_1)} \frac{1}{[g(z)-g_0]^{\frac{1}{4}}}. \tag{6.9.22}$$

Hence

$$G \sim AD(z)\,\varpi^{-\frac{1}{2}} \delta^{\frac{3}{4}} T^{-\frac{5}{4}} H(T) e^{-\xi} \quad (z_0=0,\ z>0), \tag{6.9.23}$$

and similarly

$$G \sim AD(z_0)\,D(z)\,\varpi^{-\frac{1}{2}}\delta^{\frac{1}{2}}T^{-1}H(T)\,\mathrm{e}^{-\xi} \quad (z_0 > 0,\ z > 0). \quad (6.9.24)$$

The expression $D(z)\,\varpi^{-\frac{1}{2}}$ is a divergence factor which satisfies the transport equation associated with the diffracted fronts.

The calculations in the two-dimensional case are similar, but they need not be performed. According to (6.9.18), (6.9.8) and (6.9.9) the approximate form of $\bar{G}$ in the two-dimensional case can be obtained from the three-dimensional case by multiplying by

$$\left(\frac{2\pi\varpi}{\mu_j}\right)^{\frac{1}{2}} \sim \frac{(2\pi\varpi)^{\frac{1}{2}}}{g_0^{\frac{1}{4}}s^{\frac{1}{2}}}$$

and replacing ϖ by $|x|$. The corresponding factor for the approximation to the inverse Laplace transform is

$$\frac{(2\pi)^{\frac{1}{2}}6^{\frac{3}{4}}}{\alpha_1^{\frac{3}{4}}}\frac{g_0^{\frac{1}{8}}}{g_1^{\frac{1}{2}}}\delta^{-\frac{3}{4}}T^{\frac{3}{4}}.$$

Hence if

$$B = \frac{6^{\frac{1}{4}}}{4(2\pi)^{\frac{1}{2}}\alpha_1^{\frac{3}{4}}}\frac{g_1^{\frac{1}{2}}}{g_0^{\frac{5}{8}}}, \quad (6.9.25)$$

then the diffraction formulae for the two-dimensional case are

$$\left.\begin{aligned} &G \sim B\delta^{\frac{1}{4}}T^{-\frac{3}{4}}H(T)\,\mathrm{e}^{-\xi} && (z = z_0 = 0),\\ &G \sim BD(z)\,T^{-\frac{1}{2}}H(T)\,\mathrm{e}^{-\xi} && (z_0 = 0,\ z > 0),\\ &G \sim BD(z_0)\,D(z)\,\delta^{-\frac{1}{4}}T^{-\frac{1}{4}}H(T)\,\mathrm{e}^{-\xi} && (z_0 > 0,\ z > 0).\end{aligned}\right\} \quad (6.9.26)$$

APPENDIX

Asymptotic behaviour of the eigenvalues and eigenfunctions of the circular cylinder

1. *Convergence of the eigenfunction expansion.* The convergence of the eigenfunction expansion (6.3.12) of the Laplace transform of the Green's function of the circular cylinder for real positive s is a consequence of general theorems on eigenfunctions which were referred to in the text. These theorems do not hold when s is complex, as the eigenfunctions still satisfy the orthogonality conditions

$$\int_a^\infty \phi_j\phi_k\frac{\mathrm{d}r}{r} = \delta_{jk}$$

instead of the conditions usually employed in the theory of eigen-functions which would be

$$\int_a^\infty \phi_j \phi_k^* \frac{dr}{r} = \delta_{jk},$$

where ϕ_k^* is the complex conjugate of ϕ_k. It is therefore necessary to investigate the convergence properties of the eigenfunction expansion in this case directly. For this purpose one must consider $K_{i\mu}(z)$ with $z = sa/c$ or $z = sr/c$ when $\mathcal{R}z > 0$, $\mathcal{R}\mu > 0$ and $|\mu/z|$ is large. By definition,

$$K_{i\mu}(z) = \frac{\pi}{2i\sinh\mu}\{I_{-i\mu}(z) - I_{i\mu}(z)\}, \qquad (6.A.1)$$

$$I_{i\mu}(z) = \sum_{n=1}^\infty \frac{(\tfrac12 z)^{i\mu+2n}}{n!\,(i\mu+n)!}, \qquad (6.A.2)$$

where $$(\tfrac12 z)^{i\mu+2n} = (\tfrac12 z)^{2n}\exp\{i\mu\log\tfrac12 z\},$$

$\log(\tfrac12 z)$ being the principal value which is real and positive when z is real and positive. Since

$$(i\mu)!\,(-i\mu)! = \frac{\pi\mu}{\sinh\pi\mu},$$

we have therefore for large μ/z

$$K_{i\mu}(z) = \frac{1}{2i\mu}\{(i\mu)!\,(\tfrac12 z)^{-i\mu} - (-i\mu)!\,(\tfrac12 z)^{i\mu}\}\left\{1 + O\!\left(\frac{1}{\mu}\right)\right\}.$$

By Stirling's formula,

$$(i\mu)! = (2\pi i\mu)^{\frac12}\exp\left\{i\mu\log\frac{i\mu}{e}\right\}\left\{1 + O\!\left(\frac{1}{\mu}\right)\right\},$$

provided that $-\tfrac12\pi + \delta \leqslant \arg\mu \leqslant \tfrac12\pi - \delta$, $\delta > 0$. Since then

$$\delta \leqslant \arg(i\mu) \leqslant \pi - \delta$$

one has $\log(i\mu) = \log\mu + \tfrac12 i\pi$ so that

$$(i\mu)! = (2\pi\mu)^{\frac12}\exp\left\{i\mu\log\frac{\mu}{e} - \tfrac12\pi\mu + \tfrac14 i\pi\right\}\left\{1 + O\!\left(\frac{1}{\mu}\right)\right\}.$$

Similarly since $\log(-i\mu) = \log(i\mu) - i\pi = \log\mu - \tfrac12 i\pi,$

$$(-i\mu)! = (2\pi\mu)^{\frac12}\exp\left\{-i\mu\log\frac{\mu}{e} - \tfrac12\pi\mu - \tfrac14 i\pi\right\}\left\{1 + O\!\left(\frac{1}{\mu}\right)\right\}.$$

Hence $$K_{i\mu}(z) = \left(\frac{2\pi}{\mu}\right)^{\frac12} e^{-\frac12\pi\mu}\left\{\sin\left(\mu\log\frac{2\mu}{ez} + \pi\right) + O\!\left(\frac{1}{\mu}\right)\right\}, \qquad (6.A.3)$$

$$K'_{i\mu}(z) = -\frac{(2\pi\mu)^{\frac12}}{z}e^{-\frac12\pi\mu}\left\{\cos\left(\mu\log\frac{2\mu}{ez} + \tfrac14\pi\right) + O\!\left(\frac{1}{\mu}\right)\right\}. \qquad (6.A.4)$$

The eigenvalues are the zeros of $K_{1\mu}(sa/c)$. It is easily shown from (6.A.4) that

$$\mu_j = \nu_j + O\left(\frac{1}{j}\right), \quad \nu_j \log\frac{2\nu_j c}{eas} = (N+j+\tfrac{1}{4})\pi \tag{6.A.5}$$

when j is large, where N is a fixed integer. Hence by (6.A.3) and (6.A.4)

$$K_{1\mu_j}\left(\frac{sa}{c}\right) = (-1)^{N+j}\left(\frac{2\pi}{\nu_j}\right)^{\frac{1}{2}} e^{-\frac{1}{2}\pi\nu_j}\left\{1 + O\left(\frac{1}{j}\right)\right\},$$

$$\left[\frac{\partial}{\partial\mu}K_{1\mu_j}\left(\frac{sa}{c}\right)\right]_{\mu=\mu_j} = (-1)^{N+j}\frac{(2\pi\nu_j)^{\frac{1}{2}}c}{as}\left(1+\log\frac{2\nu_j c}{eas}\right)e^{-\frac{1}{2}\pi\nu_j}\left\{1+O\left(\frac{1}{j}\right)\right\},$$

$$K_{1\mu_j}\left(\frac{sr}{c}\right) = (-1)^{N+j}\left(\frac{2\pi}{\nu_j}\right)^{\frac{1}{2}} e^{-\frac{1}{2}\pi\nu_j}\left\{\cos\left(\nu_j\log\frac{r}{a}\right)+O\left(\frac{1}{j}\right)\right\}.$$

The general term of the series (6.3.12) is therefore

$$\bar{F}_j = \frac{[\cos(\nu_j\log r_0/a)\cos(\nu_j\log r/a) + O(1/j)]}{\nu_j + (N+j+\frac{1}{4})\pi}e^{-\nu_j|\theta|},$$

and hence for all sufficiently large j

$$|\bar{F}_j| < 2\exp\left\{-|\theta|\mathscr{R}(\nu_j) + \mathscr{I}(\nu_j)\log\frac{rr_0}{a^2}\right\}. \tag{6.A.6}$$

Now if we put

$$\nu_j = k\rho_j e^{i\psi_j}, \quad s = |s|e^{i\alpha}, \quad k = \frac{ea|s|}{2c},$$

then ρ_j and ψ_j satisfy the equations

$$\rho_j[\cos\psi_j\log\rho_j + (\alpha-\psi_j)\sin\psi_j] = \frac{\pi}{k}(N+j+\tfrac{1}{4}),$$

$$\sin\psi_j\log\rho_j - (\alpha_j-\psi_j)\cos\psi_j = 0,$$

which follow from (6.A.5). Obviously $\rho_j\to\infty$ as $j\to\infty$; the second equation therefore implies that $\psi_j\to 0$; in fact, $\psi_j \sim \alpha/\log\rho_j$. Hence for all sufficiently large j

$$\mathscr{I}(\nu_j)\log\frac{rr_0}{a^2} = k\rho_j\sin\psi_j\log\frac{rr_0}{a^2} < \tfrac{1}{4}k\rho_j|\theta|\cos\psi_j = \tfrac{1}{4}|\theta|\mathscr{R}(\nu_j),$$

so that by (6.A.6)

$$|\bar{F}_j| < 2\exp(-\tfrac{3}{4}k\rho_j|\theta|\cos\psi_j),$$

and hence as $\cos\psi_j\to 1$

$$|\bar{F}| < 2\exp(-\tfrac{1}{2}k|\theta|\rho_j).$$

Now

$$\rho_j^2 - 1 > (\rho_j+1)\log\rho_j = \frac{\pi(N+j+\frac{1}{4})}{k\cos\psi_j} + [1-(\alpha-\psi_j)\tan\psi_j]\rho_j$$

$$\geqslant \frac{\pi}{k}(N+j+\tfrac{1}{4}),$$

so that for all sufficiently large j

$$|F_j| = O\{\exp[-\tfrac{1}{2}(\pi kj)^{\frac{1}{2}}|\theta|]\},$$

and since
$$\sum_{j=1}^{\infty} \exp[-\tfrac{1}{2}|\theta|(\pi kj)^{\frac{1}{2}}]$$

converges it follows that the eigenfunction expansion converges for $\mathscr{R}s \geqslant 0$ if $\theta \neq 0$.

2. *Approximations for $K_{1\mu}(sr/c)$ valid when $\mu/s \sim 1$.* We now turn to the case where both μ and s are large and of the same order of magnitude. The function

$$u(z) = K_{1\mu}(\mu z) \tag{6.A.7}$$

satisfies the differential equation

$$u''(z) + \frac{1}{z}u'(z) + \mu^2\left(\frac{1}{z^2} - 1\right)u(z) = 0. \tag{6.A.8}$$

Let us suppose that μ and z are real. There are two boundary conditions, namely, $u'(sa/\mu c) = 0$ and $\lim_{z\to\infty} u(z) = 0$. They can only be satisfied simultaneously if μ is an eigenvalue. It can be shown that u is non-oscillatory in $z > 1$ so that we shall have $sa/\mu c < 1$. Now for large μ,

$$u \sim L(\mu)\,e^{-\mu\zeta}\left(A_0 + \frac{A_1}{\mu} + \dots\right), \quad \zeta' = \left(1 - \frac{1}{z^2}\right)^{\frac{1}{2}} \quad (z > 1),$$

$$u \sim M(\mu)\,e^{i\mu\eta}\left(B_0 + \frac{B_1}{\mu} + \dots\right) + N(\mu)\,e^{-i\mu\eta}\left(C_0 + \frac{C_1}{\mu} + \dots\right),$$

$$\eta' = \left(\frac{1}{z^2} - 1\right)^{\frac{1}{2}} \quad (z < 1),$$

where the A_n, B_n and C_n are functions of z but not of μ that satisfy certain recurrence relations and L, M and N are arbitrary. Both these asymptotic expansions break down as $z \to 1$. They are therefore inadequate for our purpose, and we must use a more elaborate method of asymptotic approximation which yields results uniformly valid in the whole interval $(sa/\mu c, \infty)$. Problems of this type are frequently met with in mathematical physics, and have been discussed by several authors; the first rigorous and systematic treatment is due to Langer.† For our purpose Langer's

† Langer (1931, 1932) where the case of complex parameters is considered. A convenient summary of Langer's method will be found in Lighthill (1950). A systematic recent treatment of the subject is due to F. W. J. Olver (1955), where a review of the earlier results and full references will be found.

method can be formulated as follows. Let us define a new independent variable Z by

$$\left.\begin{aligned} \tfrac{2}{3}Z^{\frac{3}{2}} &= \int_1^z \left(1-\frac{1}{z'^2}\right)^{\frac{1}{2}} dz' = (z^2-1)^{\frac{1}{2}} - \cos^{-1}\frac{1}{z} \quad (z \geqslant 1), \\ \tfrac{2}{3}(-Z)^{\frac{3}{2}} &= \int_z^1 \left(\frac{1}{z'^2}-1\right)^{\frac{1}{2}} dz' = \cosh^{-1}\frac{1}{z} - (1-z^2)^{\frac{1}{2}} \quad (0<z\leqslant 1), \end{aligned}\right\} \tag{6.A.9}$$

where it is assumed that z is real and that $0 \leqslant \cos^{-1}(1/z) \leqslant \frac{1}{2}\pi$. Then $Z \geqslant 0$ for $z \geqslant 1$ and $Z \leqslant 0$ for $0 < z \leqslant 1$, and

$$Z\left(\frac{dZ}{dz}\right)^2 = 1 - \frac{1}{z^2}, \quad \frac{dZ}{dz} > 0. \tag{6.A.10}$$

When $z-1$ is small then $Z \sim 2^{\frac{1}{3}}(z-1)$, so that Z is a regular function of z in $(0, \infty)$ that is steadily increasing. If now

$$u(z) = (zZ')^{-\frac{1}{2}} v(Z) = \left(\frac{Z}{z^2-1}\right)^{\frac{1}{4}} v(Z), \tag{6.A.11}$$

then it follows from (6.A.8) after some calculation that

$$v''(Z) - [\mu^2 Z + q(Z)]\, v(Z) = 0,$$

where q is continuous at $Z=0$. From this one can deduce that v differs by a small term (of order $\mu^{-\frac{2}{3}}$) from the solution of the equation obtained by omitting q,

$$w''(Z) - \mu^2 Z w(Z) = 0.$$

The solution of this equation that tends to zero as $Z \to \infty$ is

$$w = L\,\mathrm{Ai}\,(\mu^{\frac{2}{3}}Z),$$

where L is independent of Z and Ai denotes the *Airy function*†

$$\mathrm{Ai}\,(\alpha) = \frac{1}{\pi}\int_0^\infty \cos\left(\tfrac{1}{3}\sigma^3 + \alpha\sigma\right) d\sigma = \frac{\alpha^{\frac{1}{2}}}{3^{\frac{1}{2}}\pi} K_{\frac{1}{3}}\left(\tfrac{2}{3}\alpha^{\frac{3}{2}}\right), \tag{6.A.12}$$

which satisfies the differential equation

$$\mathrm{Ai}''(\alpha) = \alpha\,\mathrm{Ai}\,(\alpha). \tag{6.A.13}$$

Hence
$$u(z) = K_{1\mu}(\mu z) \sim L\left(\frac{Z}{z^2-1}\right)^{\frac{1}{4}} \mathrm{Ai}\,(\mu^{\frac{2}{3}}Z). \tag{6.A.14}$$

To determine L we can compare the asymptotic form of this approximation for small z with the formulae of the preceding section. For small z (6.A.9) gives

$$\tfrac{2}{3}(-Z)^{\frac{3}{2}} \sim \log\frac{2}{ze},$$

† This definition of the Airy function is due to Sir Harold Jeffreys. For a summary of the properties of the Airy function, see J. C. P. Miller (1946).

and since for large negative argument

$$\mathrm{Ai}\,(\alpha) \sim \frac{\mathrm{I}}{\pi^{\frac{1}{2}}\,|\,\alpha\,|^{\frac{1}{4}}} \sin\{\tfrac{2}{3}(-\alpha)^{\frac{3}{2}} + \tfrac{1}{4}\pi\}, \qquad (6.\mathrm{A}.15)$$

we have

$$u(z) \sim \frac{L}{\pi^{\frac{1}{2}}\mu^{\frac{1}{6}}} \sin\left\{\mu \log \frac{2}{\mathrm{e}z} + \tfrac{1}{4}\pi\right\}.$$

On the other hand, it follows from (6.A.3) that

$$K_{i\mu}(\mu z) \sim \left(\frac{2\pi}{\mu}\right)^{\frac{1}{2}} \mathrm{e}^{-\frac{1}{2}\pi\mu} \sin\left(\mu \log \frac{2}{\mathrm{e}z} + \tfrac{1}{4}\pi\right). \qquad (6.\mathrm{A}.16)$$

Hence

$$L = \pi\left(\frac{2}{\mu}\right)^{\frac{1}{2}} \mu^{\frac{1}{6}}\,\mathrm{e}^{-\frac{1}{2}\pi\mu}. \qquad (6.\mathrm{A}.17)$$

If we finally put $z = sr/\mu c$ and $\zeta = \mu^{\frac{2}{3}}Z$ so that

$$\left.\begin{array}{ll} \tfrac{2}{3}\zeta^{\frac{3}{2}} = \left(\dfrac{s^2r^2}{c^2} - \mu^2\right)^{\frac{1}{2}} - \mu \cosh^{-1}\dfrac{\mu c}{sr} & (r \geqslant \mu c/s), \\[4mm] \tfrac{2}{3}(-\zeta)^{\frac{3}{2}} = \mu \cos^{-1}\dfrac{\mu c}{sr} - \left(\mu^2 - \dfrac{s^2r^2}{c^2}\right)^{\frac{1}{2}} & (r \leqslant \mu c/s), \end{array}\right\} \qquad (6.\mathrm{A}.18)$$

then

$$K_{i\mu}\left(\frac{sr}{c}\right) \sim 2^{\frac{1}{2}}\pi\,\mathrm{e}^{-\frac{1}{2}\pi\mu}\left(\frac{\zeta c^2}{s^2r^2 - \mu^2c^2}\right)^{\frac{1}{4}} \mathrm{Ai}\,(\zeta). \qquad (6.\mathrm{A}.19)$$

It can be shown that the error is of order $s^{-\frac{2}{3}}\,\mathrm{e}^{-\frac{1}{2}\pi\mu}$.

3. *The eigenvalues.* We can now determine the behaviour of an eigenvalue μ_j of fixed order as $s \to \infty$. An asymptotic approximation to $K'_{i\mu}(sa/c)$ can be obtained from (6.A.19) by differentiation, but as it is rather complicated it is better to argue as follows. The zeros μ_j of $K'_{i\mu}(sa/c)$ interlace those of $K_{i\mu}(sa/c)$.† As $s \to \infty$, the value of ζ associated with a zero of $K_{i\mu}$ tends to a zero of $\mathrm{Ai}\,(\zeta)$, that is to say, to a constant. Hence the ζ associated with a μ_j remains bounded as $s \to \infty$. By (6.A.18) this implies that $(\mu_j c/sa) \to 1$ as $s \to \infty$. Let

$$\eta = -\zeta(a), \quad \tfrac{2}{3}\eta^{\frac{3}{2}} = \mu \cosh^{-1}\frac{\mu c}{sa} - \left(\mu^2 - \frac{s^2a^2}{c^2}\right)^{\frac{1}{2}}. \qquad (6.\mathrm{A}.20)$$

If we put for the moment

$$y = \left(\frac{c}{sa}\right)^{\frac{2}{3}}\eta, \quad x = \frac{\mu c}{sa},$$

---

† Let $s_j(\mu)$ be the zeros of $K_{i\mu}(sa/c)$ and $s'_j(\mu)$ those of $K'_{i\mu}(sa/c)$, $K_{i\mu}$ and $K'_{i\mu}$ being considered as functions of s. Then $(\mu c/a) > s'_1 > s_1 > s'_2 > s_2 > \dots$, since the maxima and minima of $K_{i\mu}$ separate its zeros and $K_{i\mu}$ is non-oscillatory for $s > \mu c/a$. It follows from the Sturm oscillation theorem that $s_j(\mu)$ and $s'_j(\mu)$ are increasing functions of μ so that they can be inverted as $\mu = \nu_j(s)$, $\mu = \mu_j(s)$ respectively. It is then easy to see, for instance, by drawing the graphs of s_j and s'_j in an $s\mu$-plane, that $\mu_1(s) < \nu_1(s) < \mu_2(s) < \nu_2(s) < \dots$.

then
$$\tfrac{2}{3}y^{\frac{3}{2}} = x\cosh^{-1}x - (x^2-1)^{\frac{1}{2}},$$

whence
$$y^{\frac{1}{2}}\frac{dy}{dx} = \cosh^{-1}x = 2^{\frac{1}{2}}(x-1)^{\frac{1}{2}}\left\{1 - \frac{x-1}{12} + \ldots\right\},$$
$$\tfrac{2}{3}y^{\frac{3}{2}} = \tfrac{2}{3}2^{\frac{3}{2}}(x-1)^{\frac{3}{2}}\left\{1 - \frac{x-1}{20} + \ldots\right\},$$
$$y = 2^{\frac{1}{3}}(x-1)\left\{1 - \frac{x-1}{30} + \ldots\right\},$$
$$x - 1 = 2^{-\frac{1}{3}}y\left\{1 + \frac{2^{-\frac{1}{3}}y}{30} + \ldots\right\}.$$

Hence $(\mu c/sa) - 1$ is a series in ascending powers of $\eta s^{-\frac{2}{3}}$. We therefore put
$$\mu = \frac{sa}{c} + \beta\left(\frac{sa}{c}\right)^{\frac{1}{3}} + O(s^{-\frac{1}{3}}). \tag{6.A.21}$$

Then
$$\left.\begin{aligned} \eta &= 2^{\frac{1}{3}}\beta + O(s^{-\frac{2}{3}}), \\ \mu^2 - \frac{s^2a^2}{c^2} &= 2\beta\left(\frac{sa}{c}\right)^{\frac{4}{3}} + O(s^{\frac{2}{3}}). \end{aligned}\right\} \tag{6.A.22}$$

Hence by (6.A.19)
$$K_{1\mu}\left(\frac{sa}{c}\right) \sim 2^{\frac{1}{2}}\pi\, e^{-\frac{1}{2}\pi\mu}\left(\frac{\eta c^2}{\mu^2c^2 - s^2a^2}\right)^{\frac{1}{4}}\mathrm{Ai}\,(-\eta) \sim \pi\left(\frac{2c}{sa}\right)^{\frac{1}{3}} e^{-\frac{1}{2}\pi\mu}\,\mathrm{Ai}\,(-2^{\frac{1}{3}}\beta). \tag{6.A.23}$$

The approximation to $K_{1\mu}$ obtained by differentiating (6.A.19) is
$$K'_{1\mu}\left(\frac{sa}{c}\right) \sim 2^{\frac{1}{2}}\pi\, e^{-\frac{1}{2}\pi\mu}\left\{\frac{c}{sa}\left(\frac{c^2\mu^2 - s^2a^2}{c^2\eta}\right)^{\frac{1}{4}}\mathrm{Ai}'\,(-\eta)\right.$$
$$\left. + \frac{c}{4sa}\eta^{-\frac{5}{4}}\left(\mu^2 - \frac{s^2a^2}{c^2}\right)^{-\frac{5}{4}}\left[\frac{2s^2a^2}{c^2}\eta^{\frac{3}{2}} - \left(\mu^2 - \frac{s^2a^2}{c^2}\right)^{\frac{3}{2}}\right]\mathrm{Ai}\,(-\eta)\right\}.$$

When (6.A.22) is substituted in this the factor of $\mathrm{Ai}\,(-\eta)$ is found to be of higher order than that of $\mathrm{Ai}'\,(-\eta)$ so that one is left with
$$K'_{1\mu}\left(\frac{sa}{c}\right) \sim \pi\left(\frac{2c}{sa}\right)^{\frac{2}{3}} e^{-\frac{1}{2}\pi\mu}\,\mathrm{Ai}'\,(-2^{\frac{1}{3}}\beta). \tag{6.A.24}$$

Hence finally by (6.A.21)
$$\mu_j = \frac{sa}{c} + \alpha_j\left(\frac{sa}{2c}\right)^{\frac{1}{3}} + O(s^{-\frac{1}{3}}), \tag{6.A.25}$$

where $-\alpha_1, -\alpha_2, \ldots$ are the zeros of $\mathrm{Ai}'\,(\alpha)$ arranged in descending order. They are all negative, that is to say, $\alpha_j > 0$.

Our argument applies only when s is real. But it can be extended to complex s since in fact Langer's method then also yields the approximation (6.A.19), provided that ζ is defined properly.

4. *Asymptotic formulae for the* $\overline{F}_j$. The general term of the eigenfunction of the Green's function of the circular cylinder on the Riemann surface $\mathscr{R}$ is by (6.3.12)

$$\overline{F}_j=\frac{c}{sa}\frac{K_{\mathrm{i}\mu_j}\left(\frac{sr_0}{c}\right)K_{\mathrm{i}\mu_j}\left(\frac{sr}{c}\right)}{K_{\mathrm{i}\mu_j}\left(\frac{sa}{c}\right)\left[\frac{\partial}{\partial\mu}K'_{\mathrm{i}\mu}\left(\frac{sa}{c}\right)\right]_{\mu=\mu_j}}\,\mathrm{e}^{-\mu_j|\theta|}. \tag{6.A.26}$$

From (6.A.23) and (6.A.25) we have at once

$$K_{\mathrm{i}\mu_j}\left(\frac{sa}{c}\right)\sim\pi\left(\frac{2c}{sa}\right)^{\frac{1}{3}}\mathrm{e}^{-\frac{1}{2}\pi\nu_j}\,\mathrm{Ai}\,(-\alpha_j), \tag{6.A.27}$$

where

$$\nu_j=\frac{sa}{c}+\alpha_j\left(\frac{sa}{2c}\right)^{\frac{1}{3}}. \tag{6.A.28}$$

Again, since by (6.A.24)

$$K'_{\mathrm{i}\nu}\left(\frac{sa}{c}\right)\sim\pi\left(\frac{2c}{sa}\right)^{\frac{2}{3}}\mathrm{e}^{-\frac{1}{2}\pi\nu}\,\mathrm{Ai}'\,(-\alpha),\quad \nu=\frac{sa}{c}+\alpha\left(\frac{sa}{2c}\right)^{\frac{1}{3}},$$

it follows that

$$\begin{aligned}\left[\frac{\partial}{\partial\alpha}K'_{\mathrm{i}\nu}\left(\frac{sa}{c}\right)\right]_{\alpha=\alpha_j}&=\left(\frac{sa}{2c}\right)^{\frac{1}{3}}\left[\frac{\partial}{\partial\nu}K'_{\mathrm{i}\nu}\left(\frac{sa}{c}\right)\right]_{\alpha=\alpha_j}\\&\sim\left(\frac{sa}{2c}\right)^{\frac{1}{3}}\left[\frac{\partial}{\partial\mu}K'_{\mathrm{i}\mu}\left(\frac{sa}{c}\right)\right]_{\mu=\mu_j}\\&\sim-\pi\left(\frac{2c}{sa}\right)^{\frac{2}{3}}\mathrm{e}^{-\frac{1}{2}\pi\nu_j}\,\mathrm{Ai}''(-\alpha_j),\end{aligned}$$

and as $\mathrm{Ai}''(-\alpha_j)=-\alpha_j\,\mathrm{Ai}\,(-\alpha_j)$ by (6.A.13),

$$\left[\frac{\partial}{\partial\mu}K'_{\mathrm{i}\mu}\left(\frac{sa}{c}\right)\right]_{\mu=\mu_j}\sim 2\pi\frac{c}{sa}\alpha_j\,\mathrm{Ai}\,(-\alpha_j)\,\mathrm{e}^{-\frac{1}{2}\pi\nu_j}. \tag{6.A.29}$$

In order to deal with $K_{\mathrm{i}\mu_j}(sr/c)$ when $r>a$ we note first that by (6.A.25)

$$\frac{c\mu_j}{sr}=\frac{a}{r}+O(s^{-\frac{2}{3}})<1,$$

when s is sufficiently large. Hence we must use (6.A.19) with large positive ζ. The asymptotic formula for the Airy function,

$$\mathrm{Ai}\,(\zeta)\sim\frac{1}{2\pi^{\frac{1}{2}}\zeta^{\frac{1}{4}}}\exp\left(-\tfrac{2}{3}\zeta^{\frac{3}{2}}\right), \tag{6.A.30}$$

then gives
$$K_{1\mu_j}\left(\frac{sr}{c}\right) \sim \left(\frac{\pi}{2}\right)^{\frac{1}{2}} \frac{e^{-\frac{1}{2}\pi\nu_j}}{\left(\frac{s^2r^2}{c^2}-\mu_j^2\right)^{\frac{1}{4}}} \exp\left(-\tfrac{2}{3}\zeta_j^{\frac{3}{2}}\right),$$
where
$$\tfrac{2}{3}\zeta_j^{\frac{3}{2}} = \left(\frac{s^2r^2}{c^2}-\mu_j^2\right)^{\frac{1}{2}} - \mu_j \cos^{-1}\frac{\mu_j c}{sr}.$$

Substituting (6.A.25) here one finds after a simple calculation that
$$\tfrac{2}{3}\zeta_j^{\frac{3}{2}} = \frac{s}{c}\left[(r^2-a^2)^{\frac{1}{2}} - a\cos^{-1}\frac{a}{r}\right] - \left(\frac{sa}{2c}\right)^{\frac{1}{3}} \alpha_j \cos^{-1}\frac{a}{r} + O(s^{-\frac{1}{3}}).$$
Hence
$$K_{1\mu_j}\left(\frac{sr}{c}\right) \sim \left(\frac{\pi c}{2s}\right)^{\frac{1}{2}} \frac{e^{-\frac{1}{2}\pi\nu_j}}{(r^2-a^2)^{\frac{1}{4}}} \exp\left\{-\frac{s}{c}\left[(r^2-a^2)^{\frac{1}{2}} - a\cos^{-1}\frac{a}{r}\right] - \alpha_j\left(\frac{sa}{2c}\right)^{\frac{1}{3}} \cos^{-1}\frac{a}{r}\right\}. \quad (6.A.31)$$

We must now substitute these approximations in (6.A.26). Let us begin with the case $r > a$, $r_0 > a$. Since the $\exp(-\frac{1}{2}\pi\nu_j)$ cancel, the argument of the exponential factor is
$$-\frac{s}{c}\left[(r^2-a^2)^{\frac{1}{2}} + (r_0^2-a^2)^{\frac{1}{2}} + a\left(|\theta| - \cos^{-1}\frac{a}{r} - \cos^{-1}\frac{a}{r_0}\right)\right] - \alpha_j\left(\frac{sa}{2c}\right)^{\frac{1}{3}}\left[|\theta| - \cos^{-1}\frac{a}{r} - \cos^{-1}\frac{a}{r_0}\right] = -s\tau - \alpha_j\left(\frac{sa}{2c}\right)^{\frac{1}{3}}\delta.$$

The exponential is to be multiplied by
$$\frac{c}{sa}\,\frac{\dfrac{\pi c}{2s}\,\dfrac{1}{(r^2-a^2)^{\frac{1}{4}}(r_0^2-a^2)^{\frac{1}{4}}}}{\pi\left(\dfrac{2c}{sa}\right)^{\frac{1}{3}} 2\pi\dfrac{c}{sa}\,\alpha_j[\mathrm{Ai}(-\alpha_j)]^2} = \frac{a^{\frac{1}{3}}c^{-\frac{2}{3}}s^{-\frac{2}{3}}}{2^{\frac{7}{3}}\pi\alpha_j[\mathrm{Ai}(-\alpha_j)]^2(r^2-a^2)^{\frac{1}{4}}(r_0^2-a^2)^{\frac{1}{4}}}.$$

Thus the formula (6.4.1) of the text is obtained. If either $r_0 = a$ or $r = a$ then (6.A.26) simplifies; for instance
$$\bar{F}_j(a,\theta,s;r_0) = \frac{c}{sa}\,\frac{K_{1\mu_j}\left(\dfrac{sr_0}{c}\right)}{\left[\dfrac{\partial}{\partial\mu}K'_{1\mu}\left(\dfrac{sa}{c}\right)\right]_{\mu=\mu_j}} e^{-\mu_j|\theta|}.$$

There is now an exponential factor
$$\exp\left\{-\frac{s}{c}\left[(r_0^2-a^2)^{\frac{1}{2}} + a\left(|\theta| - \cos^{-1}\frac{a}{r_0}\right)\right] - \alpha_j\left(\frac{sa}{2c}\right)^{\frac{1}{3}}\left[|\theta| - \cos^{-1}\frac{a}{r_0}\right]\right\}.$$

But this is still
$$\exp\left\{-s\tau - \alpha_j\left(\frac{sa}{2c}\right)^{\frac{1}{3}}\delta\right\},$$

with τ and δ evaluated for $r=a$. The other factor is

$$\frac{c}{sa}\frac{(\pi c/2s)^{\frac{1}{2}}(r_0^2-a^2)^{-\frac{1}{4}}}{2\pi(c/sa)\,\alpha_j\,\mathrm{Ai}(-\alpha_j)}=\frac{c^{\frac{1}{2}}s^{-\frac{1}{2}}}{2^{\frac{3}{2}}\pi^{\frac{1}{2}}\alpha_j\,\mathrm{Ai}(-\alpha_j)}\frac{1}{(r_0^2-a^2)^{\frac{1}{4}}},$$

whence (6.4.2). Finally, for $r_0=r=a$,

$$\begin{aligned}\bar{F}_j(a,\theta,s;\,a)&=\frac{c}{sa}\frac{K_{1\mu_j}(sa/c)}{\left[\dfrac{\partial}{\partial\mu}K_{1\mu}\left(\dfrac{sa}{c}\right)\right]_{\mu=\mu_j}}\mathrm{e}^{-\mu_j|\theta|}\\ &\sim\frac{c}{sa}\frac{\pi(2c/sa)^{\frac{1}{3}}\,\mathrm{Ai}(-\alpha_j)}{2\pi\dfrac{c}{sa}\alpha_j\,\mathrm{Ai}(-\alpha_j)}\mathrm{e}^{-\nu_j|\theta|}\\ &\sim\frac{1}{2^{\frac{2}{3}}\alpha_j}\left(\frac{c}{sa}\right)^{\frac{1}{3}}\exp\left\{-\frac{sa}{c}|\theta|-\alpha_j\left(\frac{sa}{2c}\right)^{\frac{1}{3}}|\theta|\right\},\end{aligned}$$

and since now $c\tau=a|\theta|$, $\delta=|\theta|$, this is (6.4.3).

BIBLIOGRAPHY

ANDERSON, D. V. (1952). Reflection of a pulse by a concave paraboloid. *J. Acoust. Soc. Amer.* **24**, 324–25.

BERRY, F. J. (1952*a*). The diffraction of sound pulses by an oscillating infinitely long strip. *Quart. J. Mech.* **5**, 324–32.

BERRY, F. J. (1952*b*). The diffraction of a sound pulse by a non-rigid semi-infinite plane screen. *Quart. J. Mech.* **5**, 333–43.

BLOKHINTZEV, D. (1946). The propagation of sound in an inhomogeneous and moving medium. *J. Acoust. Soc. Amer.* **18**, 322–34.

BREMMER, H. (1949). *Terrestrial radio waves.* New York: Elsevier.

CARSLAW, H. S. (1919). Diffraction of waves by a wedge of any angle. *Proc. Lond. Math. Soc.* (2), **18**, 291–306.

CHESTER, W. (1950*a*). The propagation of sound waves in an open-ended channel. *Phil. Mag.* (7), **41**, 11–33.

CHESTER, W. (1950*b*). The propagation of a sound pulse in the presence of a semi-infinite open-ended channel. I. *Phil. Trans.* A, **242**, 527–56; II, *Proc. Roy. Soc.* A, **203**, 33–42.

CHESTER, W. (1952). The reflection of a transient pulse by a parabolic cylinder and a paraboloid of revolution. *Quart. J. Mech.* **5**, 196–205.

COURANT, R. and FRIEDRICHS, K. O. (1948). *Supersonic flow and shock waves.* New York: Interscience.

COURANT, R. and HILBERT, D. (1937). *Methoden der Mathematischen Physik*, vol. 2. Berlin: Springer.

DOAK, P. E. (1952). The reflexion of a spherical acoustic pulse by an absorbent infinite plane and related problems. *Proc. Roy. Soc.* A, **215**, 233–54.

FOX, E. N. (1948). The diffraction of sound pulses by an infinitely long strip. *Phil. Trans.* A, **241**, 71–103.

FOX, E. N. (1949). The diffraction of two-dimensional sound pulses incident on an infinite uniform slit in a perfectly reflecting screen. *Phil. Trans.* A, **242**, 1–32.

FOX, E. N. (1952). The diffraction of a plane sound pulse incident normally on a regular grating of perfectly reflecting strips. *Proc. Roy. Soc.* A, **211**, 398–417.

FOX, L. and GOODWIN, E. T. (1953). The numerical solution of non singular integral equations. *Phil. Trans.* A, **245**, 501–34.

FRIEDLANDER, F. G. (1941). The reflexion of sound pulses by convex parabolic reflectors. *Proc. Camb. Phil. Soc.* **37**, 134–49.

FRIEDLANDER, F. G. (1942*a*). On the solutions of the wave equation with discontinuous derivatives. *Proc. Camb. Phil. Soc.* **38**, 378–82.

FRIEDLANDER, F. G. (1942*b*). On the reflexion of a spherical sound pulse by a parabolic mirror. *Proc. Camb. Phil. Soc.* **38**, 383–93.

FRIEDLANDER, F. G. (1946*a*). The diffraction of sound pulses. I–IV. *Proc. Roy. Soc.* A, **186**, 322–67.

FRIEDLANDER, F. G. (1946*b*). Simple progressive solutions of the wave equation. *Proc. Camb. Phil. Soc.* **43**, 360–73.

FRIEDLANDER, F. G. (1948). On the total reflection of plane waves. *Quart. J. Mech.* **1**, 376–84.

FRIEDLANDER, F. G. (1949). Note on the geometrical optics of diffracted wave fronts. *Proc. Camb. Phil. Soc.* **45**, 395–404.

FRIEDLANDER, F. G. (1951). On the half-plane diffraction problem. *Quart. J. Mech.* **4**, 344–57.

FRIEDLANDER, F. G. (1954). Diffraction of pulses by a circular cylinder. *Commun. Pure Appl. Math.* **7**, 705–32.

FRIEDLANDER, F. G. (1955). *Propagation of a pulse in an inhomogeneous medium.* Research Report No. EM-76, Inst. of Math. Sci., New York University.

FRIEDMAN, B. (1956). *Principles and methods of applied mathematics.* New York: Wiley.

FRIEDRICHS, K. O. and KELLER, J. B. (1955). Geometrical acoustics. II. Diffraction, reflection and refraction of a weak spherical or cylindrical shock at a plane interface. *J. Appl. Phys.* **26**, 961–6.

FRIEDRICHS, K. O. and LEWY, H. (1928). Über die Eindeutigkeit und das Abhängigkeitsgebiet der Lösungen beim Anfangswertproblem linearer hyperbolischer Differentialgleichungen. *Math. Ann.* **98**, 192–204.

GARNIR, H. G. (1952*a*). Fonctions de Green de l'opérateur métaharmonique pour les problèmes de Dirichlet et de Neumann posés dans un angle ou un dièdre. *Bull. Soc. Sci. Liège,* **21**, 119–40, 207–31.

GARNIR, H. G. (1952*b*). Sur la propagation de l'onde émise par un point dans un angle ou un dièdre parfaitement réfléchissant et le problème analogue pour la conduction de la chaleur. *Bull. Soc. Sci. Liège,* **21**, 328–44.

GARNIR, H. G. (1952*c*). Sur la transformation de Laplace des distributions. *C.R. Acad. Sci., Paris,* **234**, 583–5.

GARNIR, H. G. (1953). Propagation de l'onde émise par une source ponctuelle et instantanée dans un dioptre plan. *Bull. Soc. Sci. Liège,* **22**, 85–100, 148–62.

GARNIR, H. G. (1958). *Théorie des espaces fonctionnels hilbertiens et ses applications aux problèmes aux limites de la physique mathématique.* Basle and Stuttgart: Birkhäuser.

GERJOUY, E. (1953). Total reflection of waves from a point source. *Commun. Pure Appl. Math.* **6**, 73–91.

GUNN, J. C. (1947). Linearized supersonic aerofoil theory. *Phil. Trans.* A, **240**, 327–73.

HADAMARD, J. (1903). *Leçons sur la propagation des ondes.* Paris: Hermann.

HADAMARD, J. (1910). *Leçons sur le calcul des variations.* Paris: Hermann.

HADAMARD, J. (1923). *Lectures on Cauchy's problem.* Yale University Press.

HERGLOTZ, G. (1952). Die Green'sche Funktion der Wellengleichung für eine keilförmige Begrenzung. *Math. Ann.* **124**, 219–34.

HOBSON, E. W. (1931). *Spherical and ellipsoidal harmonics.* Cambridge University Press.

HOWARTH, L. (1948). The propagation of steady disturbances in a supersonic stream bounded on one side by a parallel subsonic stream. *Proc. Camb. Phil. Soc.* **44**, 380–90.

JOHN, F. (1954). Solutions of second order hyperbolic differential equations with constant coefficients in a domain with a plane boundary. *Commun. Pure Appl. Math.* **7**, 245–69.

JONES, D. S. (1955). Note on Witham's 'The propagation of weak spherical shocks in stars'. *Proc. Camb. Phil. Soc.* **51**, 476–85.

KAY, I. (1953). Diffraction of an arbitrary pulse by a wedge. *Commun. Pure Appl. Math.* **6**, 648–87.

KELLER, J. B. (1952). Diffraction of a shock or an electromagnetic pulse by a right-angled wedge. *J. Appl. Phys.* **23**, 1267–8.

KELLER, J. B. (1954). Geometrical acoustics. I. The theory of weak shock waves. *J. Appl. Phys.* **25**, 938–47.

KELLER, J. B. and BLANK, A. (1951). Diffraction and reflection of pulses by wedges and corners. *Commun. Pure Appl. Math.* **4**, 75–94.

KELLER, J. B. and FRIEDLANDER, F. G. (1955). Asymptotic expansions of solutions of $(\nabla^2+k^2)\,u=0$. *Commun. Pure Appl. Math.* **8**, 387–94.

KELLER, J. B. and KAY, I. (1954). Asymptotic evaluations of the field at a caustic. *J. Appl. Phys.* **25**, 876–83.

KELLER, J. B. and KELLER, H. B. (1950). Determinations of reflected and transmitted fields by geometrical optics. *J. Opt. Soc. Amer.* **40**, 48–52.

KELLER, J. B., LEWIS, R. M. and SECKLER, B. D. (1956). Asymptotic solutions of some diffraction problems. *Commun. Pure Appl. Math.* **9**, 207–66.

KLINE, M. (1954). Asymptotic solutions of linear hyperbolic partial differential equations. *J. Rat. Mech. Anal.* **3**, 315–42.

LAMB, H. (1906). On Sommerfeld's diffraction problem; and on reflexion by a parabolic mirror. *Proc. Lond. Math. Soc.* (2), **4**, 190–203.

LAMB, H. (1910). On the diffraction of a solitary wave. *Proc. Lond. Math. Soc.* (2), **8**, 422–37.

LAMB, H. (1932). *Hydrodynamics*, 6th ed. Cambridge University Press.

LANGER, R. E. (1931). On the asymptotic solutions of ordinary differential equations, with an application to the Bessel functions of large order. *Trans. Amer. Math. Soc.* **33**, 23–64.

LANGER, R. E. (1932). On the asymptotic solutions of differential equations, with an application to the Bessel functions of large complex order. *Trans. Amer. Math. Soc.* **34**, 447–80.

LAX, P. D. (1954). Weak solutions of non-linear hyperbolic equations and their numerical computation. *Commun. Pure Appl. Math.* **7**, 159–93.

LIGHTHILL, M. J. (1949). The diffraction of blast. I. *Proc. Roy. Soc.* A, **198**, 454–70.

LIGHTHILL, M. J. (1950). Reflection at a laminar boundary layer. *Quart. J. Mech.* **3**, 303–25.

LIGHTHILL, M. J. (1954). Higher approximations. In SEARS, W. R. (Editor), *General theory of high speed aerodynamics*, pp. 345–487. Princeton University Press.

LIONS, J-L. (1955). Problèmes aux limites en théorie des distributions. *Acta Math.* **94**, 13–153.

MACDONALD, H. M. (1902). *Electrical waves*, App. D. Cambridge University Press.

MARCUVITZ, N. (1951). Field representation in spherically stratified regions. *Commun. Pure Appl. Math.* **4**, 263–315.

MILLER, J. C. P. (1946). *The Airy Integral.* British Ass. Math. Tables, part vol. B. Cambridge University Press.

OBERHETTINGER, F. (1954). Diffraction of waves by a wedge. *Commun. Pure Appl. Math.* **7**, 551–65.

OLVER, F. W. J. (1955). The asymptotic solutions of linear differential equations of the second order for large values of a parameter. *Phil. Trans.* A, **247**, 307–27, 328–69.

RUBINOWICZ, A. (1920). Herstellung von Lösungen gemischter Randwertprobleme bei hyperbolischen Differentialgleichungen zweiter Ordnung durch Zusammenstückelung aus Lösungen einfacher gemischter Randwertaufgaben. *Mh. Math. Phys.* **30**, 65–79.

RUBINOWICZ, A. (1927). Zur Integration der Wellengleichung auf Riemannschen Flächen. *Math. Ann.* **96**, 648–87.

SCHWARTZ, L. (1950/51). *Théorie des distributions*, vols. **1** and **2**. Paris: Hermann.

SCHWARTZ, L. (1952). Transformation de Laplace des distributions. *Commun. Sém. Math. Univ. Lund.* (tôme consacré au Jubilé de M. Marcel Riesz), pp. 196–206.

SOMMERFELD, A. (1896). Mathematische Theorie der Diffraktion. *Math. Ann.* **47**, 317–74.

SOMMERFELD, A. (1897). Über verzweigte Potentiale in Raum. *Proc. Lond. Math. Soc.* (1), **28**, 395–429.

SOMMERFELD, A. (1901). Theoretisches über die Beugung der Röntgenstrahlen. *Z. Math. Phys.* **46**, 11–97.

TEMPLE, G. (1955). The theory of generalized functions. *Proc. Roy. Soc.* A, **228**, 175–90.

TITCHMARSH, E. C. (1946). *Eigenfunction expansions.* Oxford: Clarendon Press.

TURNER, R. D. (1956). The diffraction of a cylindrical pulse by a half-plane. *Quart. Appl. Math.* **14**, 63–73.

WARD, G. N. (1955). *Linearized theory of steady high speed flow.* Cambridge University Press.

WATSON, G. N. (1919). The diffraction of electrical waves by the earth. *Proc. Roy. Soc.* A, **95**, 83–99.

ZAREMBA, S. (1915). Sopra un teorema d'unicita relativo alla equazione delle onde sferiche. *R.C. Accad. Lincei*, **24**, 904–8.

INDEX